Nano-Modified Asphalt Binders and Mixtures to Enhance Pavement Performance

Nano-Modified Asphalt Binders and Mixtures to Enhance Pavement Performance

Editors

Luís Picado Santos
João Crucho

MDPI • Basel • Beijing • Wuhan • Barcelona • Belgrade • Manchester • Tokyo • Cluj • Tianjin

Editors
Luís Picado Santos
Universidade de Lisboa
Portugal

João Crucho
Universidade de Lisboa
Portugal

Editorial Office
MDPI
St. Alban-Anlage 66
4052 Basel, Switzerland

This is a reprint of articles from the Special Issue published online in the open access journal *Applied Sciences* (ISSN 2076-3417) (available at: https://www.mdpi.com/journal/applsci/special_issues/Nano-Modified_Asphalt).

For citation purposes, cite each article independently as indicated on the article page online and as indicated below:

LastName, A.A.; LastName, B.B.; LastName, C.C. Article Title. *Journal Name* **Year**, *Article Number*, Page Range.

ISBN 978-3-03936-710-8 (Hbk)
ISBN 978-3-03936-711-5 (PDF)

© 2020 by the authors. Articles in this book are Open Access and distributed under the Creative Commons Attribution (CC BY) license, which allows users to download, copy and build upon published articles, as long as the author and publisher are properly credited, which ensures maximum dissemination and a wider impact of our publications.

The book as a whole is distributed by MDPI under the terms and conditions of the Creative Commons license CC BY-NC-ND.

Contents

About the Editors vii

Luís Picado-Santos and João Crucho
Special Issue on Nano-Modified Asphalt Binders and Mixtures to Enhance Pavement Performance
Reprinted from: *Appl. Sci.* **2020**, *10*, 4187, doi:10.3390/app10124187 1

João Crucho, Luís Picado-Santos, José Neves and Silvino Capitão
A Review of Nanomaterials' Effect on Mechanical Performance and Aging of Asphalt Mixtures
Reprinted from: *Appl. Sci.* **2019**, *9*, 3657, doi:10.3390/app9183657 5

Wei Guo, Xuedong Guo, Wuxing Chen, Yingsong Li, Mingzhi Sun and Wenting Dai
Laboratory Assessment of Deteriorating Performance of Nano Hydrophobic Silane Silica Modified Asphalt in Spring-Thaw Season
Reprinted from: *Appl. Sci.* **2019**, *9*, 2305, doi:10.3390/app9112305 37

Wei Guo, Xuedong Guo, Jilu Li, Yingsong Li, Mingzhi Sun and Wenting Dai
Assessing the Effect of Nano Hydrophobic Silane Silica on Aggregate-Bitumen Interface Bond Strength in the Spring-Thaw Season
Reprinted from: *Appl. Sci.* **2019**, *9*, 2393, doi:10.3390/app9122393 57

Murryam Hafeez, Naveed Ahmad, Mumtaz Ahmed Kamal, Javaria Rafi, Muhammad Faizan ul Haq, Jamal, Syed Bilal Ahmed Zaidi and Muhammad Ali Nasir
Experimental Investigation into the Structural and Functional Performance of Graphene Nano-Platelet (GNP)-Doped Asphalt
Reprinted from: *Appl. Sci.* **2019**, *9*, 686, doi:10.3390/app9040686 77

Punit Singhvi, Javier J. García Mainieri, Hasan Ozer, Brajendra K. Sharma and Imad L. Al-Qadi
Effect of Chemical Composition of Bio- and Petroleum-Based Modifiers on Asphalt Binder Rheology
Reprinted from: *Appl. Sci.* **2020**, *10*, 3249, doi:10.3390/app10093249 97

Arminda Almeida, João Crucho, César Abreu and Luís Picado-Santos
An Assessment of Moisture Susceptibility and Ageing Effect on Nanoclay-Modified AC Mixtures Containing Flakes of Plastic Film Collected as Urban Waste
Reprinted from: *Appl. Sci.* **2019**, *9*, 3738, doi:10.3390/app9183738 117

Gabriela Ceccon Carlesso, Glicério Trichês, João Victor Staub de Melo, Matheus Felipe Marcon, Liseane Padilha Thives and Lídia Carolina da Luz
Evaluation of Rheological Behavior, Resistance to Permanent Deformation, and Resistance to Fatigue of Asphalt Mixtures Modified with Nanoclay and SBS Polymer
Reprinted from: *Appl. Sci.* **2019**, *9*, 2697, doi:10.3390/app9132697 131

Federico Gulisano, João Crucho, Juan Gallego and Luis Picado-Santos
Microwave Healing Performance of Asphalt Mixture Containing Electric Arc Furnace (EAF) Slag and Graphene Nanoplatelets (GNPs)
Reprinted from: *Appl. Sci.* **2020**, *10*, 1428, doi:10.3390/app10041428 147

Ming Huang and Xuejun Wen
Experimental Study on Photocatalytic Effect of Nano TiO_2 Epoxy Emulsified Asphalt Mixture
Reprinted from: *Appl. Sci.* **2019**, *9*, 2464, doi:10.3390/app9122464 163

Solomon Sackey, Dong-Eun Lee and Byung-Soo Kim
Life Cycle Assessment for the Production Phase of Nano-Silica-Modified Asphalt Mixtures
Reprinted from: *Appl. Sci.* **2019**, *9*, 1315, doi:10.3390/app9071315 . . . . **175**

About the Editors

Luís Picado Santos, PhD, is Full Professor of Transport and Infrastructures. He is President of the research centre CERIS - Civil Engineering Research and Innovation for Sustainability. Director of the Doctoral Program in Transportation Systems under the auspicious of the MIT (Massachusetts Institute of Technology)-Portugal cooperation program, which involves MIT and two Portuguese schools. He is Director of the Highways and Transport Experimental Laboratory. Since 1995, Luis has supervised 22 concluded PhD and 60 MSc dissertations. He is author of more than 300 international publications, including 52 articles on international peer reviewed journals (ISI and/or SCOPUS). His research interests are: reclaimed asphalt pavement; circular economy; asphalt mixture; recycled concrete aggregate; electric arc furnace slag; pavement management systems; urban environment road safety; dynamic traffic management.

João Crucho, PhD, is researcher in Transportation Systems at Instituto Superior Técnico, Universidade de Lisboa. Regarding pavement materials, he has conducted research in the fields of asphalt mixture ageing, asphalt binder modification and the use of alternative aggregates. As a Civil Engineer, he has developed work in the fields of infrastructure management, characterization, maintenance and rehabilitation of highways and airport pavements. He is currently participating in projects with several public and private institutions. His research interests are: pavement engineering; infrastructure management; mechanical performance; modified asphalt binders; asphalt ageing resistance; recycled aggregates.

Editorial

Special Issue on Nano-Modified Asphalt Binders and Mixtures to Enhance Pavement Performance

Luís Picado-Santos * and João Crucho *

CERIS, Instituto Superior Técnico, Universidade de Lisboa, Av. Rovisco Pais, 1049-001 Lisboa, Portugal
* Correspondence: luispicadosantos@tecnico.ulisboa.pt (L.P.-S.); joao.crucho@tecnico.ulisboa.pt (J.C.)

Received: 30 May 2020; Accepted: 3 June 2020; Published: 18 June 2020

1. Introduction

This Special Issue is dedicated to the use of nanomaterials for the modification of asphalt binders to support the analysis of the relevant properties and to determine if the modification indicated a more efficient use of asphalt mixtures' fabrication or their modification in the context of asphalt mixtures' fabrication and the improvement (or lack thereof) of these last ones to constitute effective asphalt pavement layers. All these approaches aimed to enhance performance for flexible pavements.

A total of 10 contributions were published. Four of the contributions are classified in the abovementioned first group, "Binder's modification", and five in the other group, "Asphalt mixtures' modification". The remaining contribution was a review of the effects of the modifications with nanomaterials, particularly nanosilica, nanoclays, and nanoiron, on the performance of asphalt mixtures. It could be classified in the second-mentioned group were it not for its "review" characteristics.

2. Use of Nanomaterials in the Asphalt Industry

The review published [1] described the effect of using nanosilica, nanoclays, and nanoiron to achieve better, more efficient asphalt mixtures, mechanically and in durability terms, fostering high-performance and long-lasting asphalt pavements.

Reference to several studies was already done, mostly focusing on the asphalt binder properties and its rheology, and the description of their positive findings had been the driver to the study of modified asphalt mixtures, the main focus of the review.

It could be seen that, for asphalt mixtures:

1. The modifications with nanosilica present better mechanical resistance and higher resistance to moisture damage than the other nanomaterials. The modification effect increases according to the increase in the percentage of nanomaterial used, but this could be not economically feasible.
2. The modifications with nanoclays were dependent on the type of nanoclay (raw or organic, and this last one costing the double). The use of this type of modification should be carefully defined to get excellent performance with the lowest percentage of use possible.
3. The modifications with nanoiron delivered essential improvements in the mechanical performance of the modified asphalt mixture. With a low percentage of use, if effective, this modification can be competitive.
4. The nanomaterials' modification also gave reasonable indications regarding the durability (better properties when aged) of asphalt mixtures.

These features highlight the need for a life cycle cost evaluation when addressing the use of this type of modification to establish the right balance between the construction costs, sustainability, and long-term performance of nano-modified asphalt mixtures.

3. Binders' Modification

In this group of papers, two ([2] and [3]) addressed the use of nano hydrophobic silane silica (NHSS) modification for asphalt cement and studied its behavior under freeze–thaw (F–T) aging conditions. The findings indicated that NHSS could, in certain situations, inhibit the F–T aging process of asphalt cement, but, NHSS being an inorganic material, its connection with asphalt cement is more likely to be destroyed under the F–T aging process. However [3], NHSS could increase the aggregate–bitumen interface shear strength under any working conditions, including spring thaw season. Moreover, paper [3] offers two models to evaluate the moisture damage degree and moisture damage rate of aggregate-bitumen interface shear strength.

A new material Graphene Nano-Platelets (GNPs), has been used to enhance pavements' structural and functional performances [4]. The results showed that GNPs improved not only the rutting resistance of the pavement but also its durability. The high surface area of GNPs increases the pavement's bonding strength and makes the asphalt binder stiffer. GNPs also provide nano-texture to the pavement, which enhances its skid resistance.

Finally, paper [5] evaluates the impact of modifiers' chemistry on modified binders' long-term cracking potential, meaning the recycling of reclaimed asphalt pavement material within the application of new asphalt layers. Chemical analysis indicated that the best performing modified binders had significant amounts of nitrogen in the form of amines. On the other hand, poor-performing modified binders had traces of sulfur. Additionally, modifiers with lower average molecular weights appeared to have a positive impact on the performance of aged binders. The inferences for this field of studies underlined that nanomaterials, improving aging behavior, could be a very effective asphalt cement modification, as appointed in the previous section.

4. Asphalt Mixtures' Modification

A study [6] on the moisture susceptibility of a nanoclay-modified asphalt concrete (AC) mixture containing plastic film (in flakes) collected as urban waste, evaluated with specimens subjected to an accelerated aging procedure, indicated that the combination of a nanomaterial with a by-product could be a viable solution for the recycling of plastic film, being an eco-friendly alternative to disposal in landfills.

The combination, in another study [7], of nanoclay with an SBS polymer (an elastomeric product), for the modification of asphalt mixtures, evidenced the notion that this type of mixture improved permanent deformation and leveled fatigue behavior when compared to conventional asphalt mixtures, unmodified and modified just with SBS polymer. These results could help to introduce an effective alternative for flexible pavements, where higher resistance to rutting is required.

Another exciting application was brought by the paper [8] with the study of the effect of adding Electric Arc Furnace (EAF) slag and Graphene Nanoplatelets (GNPs) on the microwave heating and healing efficiency of asphalt mixtures. The results obtained indicate that the additions of graphene and EAF slag can allow significant savings, up to 50%, on the energy required to perform a proper healing process by microwave technology, which in any case is a technology still in development.

The contribution of a type of nanomaterial, the nano-titanium dioxide nano-TiO_2, to the attenuation pollutants coming from the use of fossil fuel, an essential issue for confined environments as tunnels or underground parking places, was brought by the paper [9], investigating four influencing factors on the photocatalytic effect of the nano-TiO_2 particle sizes. The main results were that smaller particles (5 nm against 10–15 nm) and a higher dosage of nano-TiO_2 improved the elimination of hydrocarbons and nitrogen oxide significantly. The effect on the elimination of carbon oxide and carbon dioxide was not as expressive as for the other type of pollutants.

Finally, paper [10] showed the application of LCA to nano-silica-modified asphalt mixtures. It has the potential to guide decision makers on the selection of pavement modification additives to realize the benefits of using nanomaterials in pavements while avoiding potential environmental risks.

5. Final Considerations

The Guest Editors believe that this group of papers, published in this Special Issue, fosters awareness about the use of nanomaterials to modify asphalt mixtures to obtain more performant and durable flexible road pavements.

There are also other studies and applications going on, namely because there is still a route to the practical validation of the use, but this base is a robust one, especially for the researchers and practitioners interested in developing and applying these kinds of solutions.

Funding: This research received no external funding.

Acknowledgments: The guest editors wish to thank all the authors, peer reviews, and MDPI editorial staff for their valuable contributions to this Special Issue.

Conflicts of Interest: The authors declare no conflict of interest.

References

1. Crucho, J.; Picado-Santos, L.; Neves, J.; Capitão, S. A Review of Nanomaterials' Effect on Mechanical Performance and Aging of Asphalt Mixtures. *Appl. Sci.* **2019**, *9*, 3657. [CrossRef]
2. Guo, W.; Guo, X.; Chen, W.; Li, Y.; Sun, M.; Dai, W. Laboratory Assessment of Deteriorating Performance of Nano Hydrophobic Silane Silica Modified Asphalt in Spring-Thaw Season. *Appl. Sci.* **2019**, *9*, 2305. [CrossRef]
3. Guo, W.; Guo, X.; Li, J.; Li, Y.; Sun, M.; Dai, W. Assessing the Effect of Nano Hydrophobic Silane Silica on Aggregate-Bitumen Interface Bond Strength in the Spring-Thaw Season. *Appl. Sci.* **2019**, *9*, 2393. [CrossRef]
4. Hafeez, M.; Ahmad, N.; Kamal, M.; Rafi, J.; Haq, M.; Jamal; Zaidi, S.; Nasir, M. Experimental Investigation into the Structural and Functional Performance of Graphene Nano-Platelet (GNP)-Doped Asphalt. *Appl. Sci.* **2019**, *9*, 686. [CrossRef]
5. Singhvi, P.; García Mainieri, J.; Ozer, H.; Sharma, B.; Al-Qadi, I. Effect of Chemical Composition of Bio- and Petroleum-Based Modifiers on Asphalt Binder Rheology. *Appl. Sci.* **2020**, *10*, 3249. [CrossRef]
6. Almeida, A.; Crucho, J.; Abreu, C.; Picado-Santos, L. An Assessment of Moisture Susceptibility and Ageing Effect on Nanoclay-Modified AC Mixtures Containing Flakes of Plastic Film Collected as Urban Waste. *Appl. Sci.* **2019**, *9*, 3738. [CrossRef]
7. Ceccon Carlesso, G.; Trichês, G.; Staub de Melo, J.; Marcon, M.; Padilha Thives, L.; da Luz, L. Evaluation of Rheological Behavior, Resistance to Permanent Deformation, and Resistance to Fatigue of Asphalt Mixtures Modified with Nanoclay and SBS Polymer. *Appl. Sci.* **2019**, *9*, 2697. [CrossRef]
8. Gulisano, F.; Crucho, J.; Gallego, J.; Picado-Santos, L. Microwave Healing Performance of Asphalt Mixture Containing Electric Arc Furnace (EAF) Slag and Graphene Nanoplatelets (GNPs). *Appl. Sci.* **2020**, *10*, 1428. [CrossRef]
9. Huang, M.; Wen, X. Experimental Study on Photocatalytic Effect of Nano TiO2 Epoxy Emulsified Asphalt Mixture. *Appl. Sci.* **2019**, *9*, 2464. [CrossRef]
10. Sackey, S.; Lee, D.; Kim, B. Life Cycle Assessment for the Production Phase of Nano-Silica-Modified Asphalt Mixtures. *Appl. Sci.* **2019**, *9*, 1315. [CrossRef]

© 2020 by the authors. Licensee MDPI, Basel, Switzerland. This article is an open access article distributed under the terms and conditions of the Creative Commons Attribution (CC BY) license (http://creativecommons.org/licenses/by/4.0/).

Review

A Review of Nanomaterials' Effect on Mechanical Performance and Aging of Asphalt Mixtures

João Crucho [1,*], Luís Picado-Santos [1,*], José Neves [1] and Silvino Capitão [1,2]

[1] CERIS, Instituto Superior Técnico, Universidade de Lisboa, Av. Rovisco Pais, 1049-001 Lisboa, Portugal
[2] Instituto Politécnico de Coimbra, Instituto Superior de Engenharia de Coimbra, Rua Pedro Nunes, 3030-199 Coimbra, Portugal
* Correspondence: joao.crucho@tecnico.ulisboa.pt (J.C.); luispicadosantos@tecnico.ulisboa.pt (L.P.-S.); Tel.: +351-218-418-100 (J.C.); +351-218-419-715 (L.P.-S.)

Received: 3 August 2019; Accepted: 30 August 2019; Published: 4 September 2019

Abstract: This review addresses the effects of the modifications with nanomaterials, particularly nanosilica, nanoclays, and nanoiron, on the mechanical performance and aging resistance of asphalt mixtures. The desire for high-performance and long-lasting asphalt pavements significantly pushed the modification of the conventional paving asphalt binders. To cope with such demand, the use of nanomaterials for the asphalt binder modification seems promising, as with a small amount of modification an important enhancement of the asphalt mixture mechanical performance can be attained. Several studies already evaluated the effects of the modifications with nanomaterials, mostly focusing on the asphalt binder properties and rheology, and the positive findings encouraged the study of modified asphalt mixtures. This review focuses on the effects attained in the mechanical properties of the asphalt mixtures, under fresh and aged conditions. Generally, the effects of each nanomaterial were evaluated with the current state-of-art tests for the characterization of mechanical performance of asphalt mixtures, such as, permanent deformation, stiffness modulus, fatigue resistance, indirect tensile strength, and Marshall stability. Aging indicators, as the aging sensitivity, were used to evaluate the effects in the asphalt mixture's aging resistance. Finally, to present a better insight into the economic feasibility of the analyzed nanomaterials, a simple cost analysis is performed.

Keywords: modified bitumen; nanomaterials; nanosilica; nanoclay; nanoiron; asphalt mixtures; mechanical performance; aging sensitivity

1. Introduction

The asphalt binder, i.e., the bitumen, is a material widely used for road construction worldwide. Generally, the bitumen is obtained from refining crude oil and its final properties are dependent on crude oil origin and refining processes. Bitumen can be described as a thermoplastic, viscous-elastic material that behaves as a solid at low/intermediate temperatures (under 25 °C) and as a semi-solid/liquid at higher temperatures (typically above 60 °C) [1,2]. This property allows its use in road construction, where firstly, the bitumen is heated to properly mix with the aggregates and, finally, after the compaction process and cooling to ambient temperature, the bitumen will act as the binder of the aggregates. Nevertheless, the bitumen temperature sensitivity causes several problems for the asphalt pavement in service. The permanent deformation and cracking mechanics are highly related to high and low service temperatures, respectively.

While in service, the asphalt pavement has to withstand a wide range of environmental conditions and traffic loads. In many cases, the conventional penetration grade bitumen no longer ensures the desired performance over the service life, and early conservation work or reconstruction may be needed. In addition, the bitumen is a material sensitive to aging, and its properties deteriorate over the service life. The aged bitumen becomes stiffer and more brittle, thus affecting the performance of the asphalt mixture [1]. Aging effect is particularly severe in surface layers that are exposed to environmental conditions such as UV radiation, moisture, oxygen, and larger temperature change [3]. Thus, the service life of the asphalt mixture is dependent of its aging resistance [4].

Over the years, several types of additives have been studied to modify the properties of the asphalt mixtures, generally, focusing on the improvement of mechanical performance. The additives studied more frequently were adhesion improvers, fibers, rubber, to use warm mix asphalt (WMA) technology, and a wide variety of polymers [5]. In the last one or two decades, following the developments in the field of nanotechnology, the study of nanomaterials broadened and its application as asphalt mixture additive was considered.

The definition of nanomaterial encompasses a wide variety of different materials, generally, designated according to their specific properties or structures (e.g. nanoparticles, nanotubes, nanowires, nanoplatelets, nanorods, and nanoporous). Nano is a unit prefix name, represented by the symbol n, which corresponds to the submultiple 10^{-9}. Thus, the materials that have their dimensions in the nanoscale, generally 1 nm to 100 nm, are often designated as nanomaterials. The European Commission Recommendation (2011/696/EU) [6] provides a more concise definition for nanomaterial: "Natural, incidental or manufactured material containing particles, in an unbounded state or as an aggregate or as an agglomerate and where, for 50% or more of the particles in the number size distribution, one or more external dimensions is in the size range 1 nm to 100 nm". Fairly similar description is provided by the American Society for Testing and Materials (ASTM) in ASTM E2456-06 2012 [7].

The nanoscale allows the material to behave differently than its macroscopic counterpart. Such behavior can be triggered by two effects: The surface to volume ratio (specific surface area) and spatial confinement [8]. The specific surface area increases as the particle size decreases, becoming significantly large in the nanoscale. For example, in the case of a single spherical particle the surface to volume ratio is 3 mm^{-1}, 3×10^3 mm^{-1}, and 3×10^6 mm^{-1}, for the sphere radius of 1 mm, 1 µm, and 1 nm, respectively. Thus, considering the same volume unit, the use of nanoparticles instead of microparticles will allow a much larger available surface area. Nanomaterials can play a significant role in enhancing the performance of the existing materials by providing better resistance to traffic and environmental loads or mitigating incompatibility between some natural aggregates and asphalt binder, enabling more sustainable and durable pavement solutions [9].

The objective of this review is to analyze the effects of the modification with nanomaterials in the mechanical performance of the asphalt mixture. This review focuses on the modifications with nanosilica, nanoclay, and nanoiron. Firstly, the effects of the modifications in the properties and rheology of the modified bitumen are summarized, subsequently, the effects of the modifications in the mechanical performance of the asphalt mixture are analyzed as well as their contributions for aging resistance.

2. Nanomaterials

2.1. Type of Nanomaterials

Theoretically, any material can be synthetized in the nanometric scale, generically, by processing macroparticles of the respective material. The nanomaterials more studied for asphalt binder and asphalt mixture modification are types of nanosilica and of nanoclay.

Nanosilica is the term used to designate nanoparticles of silicon dioxide (SiO_2). Silicon dioxide is an inorganic material produced mainly from silica precursors, e.g. synthetized from silica fume or chemically processed from rice husk ash [10–16]. It has a molecular mass of 60.08 g/mol and the appearance of a white powder. Figure 1 presents a comparison of the volume taken by a sample of 2.50 g of nanosilica and 2.50 g of limestone filler (in the case of the filler, it is only the fraction under 63 µm). One can see that the nanoparticles occupy considerably more volume. Concerning the asphalt binder modification, the good dispersion ability and large surface area are its most interesting characteristics. Table 1 presents the properties of the nanosilica used by several researchers.

Figure 1. Mass of 2.50 g of limestone filler under 0.063 mm (**left**) and nanosilica (**right**).

Table 1. Properties of nanosilica used by several authors.

Particle Size (nm)	Specific Surface Area (m^2/g)	True Density (g/cm^3)	Bulk Density (g/cm^3)	SiO$_2$ (%)	Reference
12	175–225	2.6	–	≥99.8	[17]
20–30	130–600	2.1	–	≥99	[18]
<10	600	2.4	0.10	≥99	[19,20]
15 ± 3	160 ± 12	–	0.14	≥99.9	[21]
30	440	–	0.063	–	[22]
20–30	180–600	2.4	–	99	[23]
30	200 ± 35	–	0.03-0.06	99.8	[24]
70	64	2.2–2.6	–	–	[25]

The clays are materials that can be found abundantly in nature. Although presenting some natural variability in their constitution, such ease of access made them known materials with many applications. Currently, there are few processes to extract nanoclay from a layered clay [26,27]. Montmorillonite, is a smectite clay material derived from bentonite ore [28], is the most common natural nanomineral used by industry [29]. Majority of the clays present a layered structure, which consists of a Silica tetrahedron connected to an alumina octahedron, coordinated by oxygen atoms or hydroxyl groups, with the overall thickness of a single layer approaching one nanometer [30]. The complete separation (exfoliation) of the nanoclay layers will result in a large surface area, up to 800 m^2/g [9], as well as, very high aspect ratio, typically 100 to 1500 [31].

Generally, the natural nanoclays have hydrophilic properties. The hydrophilic behavior may cause difficulties to disperse the nanoclay homogeneously in the asphalt binder, which has organophilic properties [32]. To mitigate such a problem, the raw nanoclays can be modified by replacing the interlayer cations with quaternized ammonium or phosphonium cations, preferably with long alkyl chains, originating an organically modified or organophilic nanoclay [5], e.g. cloisite is an organically modified nanoclay which base is montmorillonite. Table 2 presents the properties of the nanoclay used by several researchers, where, in all the cases, the base of the studied nanoclays was montmorillonite. The dispersion of nanoclay in the asphalt matrix can create immiscible, intercalated, or exfoliated nanostructures [33]. In an intercalated structure, there is an expansion of the nanoclay interlayer spacing that is occupied by asphalt molecules. In an exfoliated structure, the layers of the nanoclay are exfoliated (completely separated) and the individual layers are distributed throughout the polymer matrix.

Table 2. Properties of nanoclay used by several authors.

Designation	Type	Modifier	Bulk Density (g/cm^3)	Reference
Cloisite-15A	Organophilic	Methyl, tallow, bis-2-hydroxyethyl, quaternary ammonium	0.230	[34]
Nanofil-15	Organophilic	Nanodispers layered silicate, long chain hydrocarbon	0.190	[34]
Organophilic nanoclay	Organophilic	dimethyl ammonium with two alkyl chains	–	[35]
Bentonite	Hydrophilic	–	–	[36,37]
NMN	Hydrophilic	–	0.678	[38]
PMN	Organophilic	Polysiloxane	0.251	[38]
Cloisite 15A	Organophilic	Quaternary ammonium salt	–	[39]
Nanoclay A	Organophilic	Na-activated; Dimethyl, dehydrogenated tallow, quaternary ammonium	–	[40]
Nanoclay B	Organophilic	Na-activated; Methyl, tallow, bis-2-hydroxyethyl, quaternary ammonium	–	[40]
Nanoclay C	Organophilic	Dimethyl, benzyl, Na-activated; hydrogenated tallow, quaternary ammonium	–	[40]
BT	Hydrophilic	–	–	[41,42]
OBT	Organophilic	octadecylammonium salt	–	[41,42]
Nanoclay	Organophilic	Polysiloxane	0.251	[43]

Iron nanoparticles are mostly Fe and iron oxides, such as FeO, Fe_2O_3, and Fe_3O_4. Generally, these materials are a red brown/black powder, depending of the percentage of iron oxides in its composition. The Fe nanoparticles are also commercially available in the form of zero-valent iron (ZVI), also designated zero-valent nanoiron (nZVI). ZVI can be found as a dry ferrous powder of non-valent chain presenting alkaline properties (pH from 11 to 12). Currently, ZVI has been successfully applied in groundwater remediation and wastewater treatment. Thus, the production of such nanoparticles streamlined over the last years [44–48]. Their properties such as reactivity and high specific surface may cause an important impact on the properties of the asphalt binder. Table 3 presents the properties of iron nanoparticles used by several researchers.

Table 3. Properties of nanoiron used by several authors.

Type	Particle Size (nm)	Specific Surface Area (m^2/g)	Purity (%)	Reference
Fe	50	25	>80	[36]
Fe_2O_3	38	–	–	[49]
Fe_2O_3	20–40	40–60	>98	[50]

2.2. Modification of the Asphalt Binder with Nanomaterials

In the majority of the studies found in literature, the modification of asphalt mixtures with nanomaterials is initially done at the binder level, i.e., the asphalt binder is modified with the nanomaterials, and then, the modified binder is used to produce the asphalt mixture. The optimum dosage of nanomaterial in the asphalt binder will be dependent on the type of nanomaterial, type of asphalt binder, and the methodology used, i.e., type of testing selected. Generally, the nanomaterials are blended with asphalt binder in small percentages, around 2 to 6% by mass of asphalt binder [51]. In some cases, besides the nanomaterial, a polymer modification is also done, or the binder being modified is a polymer-modified binder (PMB). Generally, for the modification of the asphalt binder with nanomaterials in laboratory one out of two methods is used: The dry blending method or the solvent blending method [52–55]. A recent review addresses more in depth the details of the polymer modification with nanoclays [56,57].

The dry blending method consists of the use of high-speed stirring to disperse the nanomaterials in the asphalt binder matrix. In this method the asphalt binder is previously heated above the softening point temperature, generally, up to a temperature equal or near to the recommended asphalt mixture mixing temperature, the nanomaterial is added, and a shear mixing effect is applied for a specific time period. As it will be additive (nanomaterial) and neat asphalt binder dependent, several trials may be needed to determine the adequate combination of rotation speed and mixing time. In addition to the high-speed shear mixing, some authors also applied sonication. Table 4 presents the dry blending configuration used by several authors.

Table 4. Dry blending configurations used by several authors.

Modification	Neat Binder	Temperature (°C)	Rotation Speed (rpm)	Duration (min.)	Reference
2%, 4% SiO_2	PG 76 PM	160	1500	60	[58]
3% OMMT	PG 58-22	150	5000	100	[35]
4% SiO_2; 4% ZVI; 4% BT	35/50	160	2000	60	[25]
0.5%, 1% CNT; 3%, 6% NC	PG 58-22	150	1550	90	[59][1]
2% CNT	PG 58-22	150	5000	100	[60]
3%, 5%, 7% NC; 3%, 5%, 7% NSF; 3%, 5%, 7% NSH	PG 52 (50/70)	145	1500	60	[27]
0.5%, 1% CNT; 3%, 6% NC	70/100 50/70	150	1550	90	[61][1]
2%, 4% OMMT	PG 64-28	160	2500	180	[55]
2%, 4% NMN; 2%, 4% PMN	PG 58-34	130	4000	120	[38]
2% to 8% SiO_2	60/70	135	4000	120	[19,20]
0.5% to 5% SiO_2	70/100	160	4000	60	[62]
1.5% OMMT	PG 58-10	180	4000	45	[39]
5%, 10% CBNP	PG 58-22	158	2800	45	[63]
4% SiO_2; 4% TiO_2; 4% $CaCO_3$	60/70	160	6000	60	[64]

[1] Using sonication.

The use of excessive rotational speed and prolonged mixing times can cause an undesired accelerated oxidation and consequent aging of the asphalt binder. The geometry of the mixing shaft head can also play an important role. The use of the most common shaft head geometries (such as blades, anchor, propeller, and Rushton) at high rotation speed can easily induce vortex effects that will potentiate the entrapment of air bubbles in the asphalt matrix. This effect, aggravated by the fact that the mixing occurs at high temperatures, may promote a significant premature oxidation of the asphalt. To mitigate this effect, the use of head geometries, such as the Jiffy head, that prevent vortex formation can be preferable. Other possibility to eliminate undesired oxidation could be to carry out the process of asphalt modification under controlled atmosphere conditions, for example using a nitrogen atmosphere furnace.

In the solvent blending method, the nanomaterial is initially dispersed in a compatible solvent (for example toluene or kerosene) that later will be mixed with the neat asphalt binder under medium to high temperature applying low speed stirring. The mixing process finishes when the evaporation of the solvent is complete. Some authors also applied sonication during the stirring time. The description of the solvent blending configurations can be found in the respective studies [65–67].

Due to several advantages, the dry blending method is the most widely used. Compared to the solvent blending method, the most important advantages of the dry blending method are that it is cheaper and easier to implement and does not require the use of high amount of solvents. For either method, the evaluation of the final modified binder properties will reveal if a homogeneous blend was achieved. To perform this evaluation, a procedure similar to the one stated by the CEN specification EN 13399—determination of storage stability of modified bitumen [68] or ASTM D5892-00 [69] can be used [63,70]. As the used dosages of nanomaterials are typically low, under 6% of the mass of asphalt binder, and the individual mass of the nanoparticles is very small, sedimentation problems were not reported.

There are some concerns regarding the safety of nanoparticles and the associated potential health risks. Because of their nanoscale dimensions, they can easily pass through biological systems, such as human skin and cell membranes and accumulate in undesirable locations up to toxic levels [71]. Grassian et al. [72] found that the inhalation of nano TiO_2 at 8.8 mg/m^3 concentration caused lung inflammation. Although some studies were already conducted, there is still a big uncertainty regarding the effects of engineered nanomaterials on environment and human health. Recently, due to the proliferation of nanotechnology in several industries, particularly food additives and packaging, more publications are addressing this topic [30,71,73–75]. The production of nanomaterials and the asphalt binder modification are important phases where exposure can be significant. The most likely routes for exposure are inhalation, ocular, and dermal adsorption [75]. Unless other information is given by the nanomaterials' suppliers, the manipulation and handling of such materials should be assumed as a potential hazard, thus safety handling protocols should be implemented accordingly. Crucho [36] modified asphalt binder in laboratory using a fume hood cabinet and individual protections: Nitrile gloves at least 0.5 mm thick, mask for eye protection, breathing mask with particle-filter FFP3, and protection suit (Tychen C—category III).

3. Effect of the Modification with Nanomaterials in the Asphalt Binder

Some studies about the use of nanomaterials in asphalt binders have already been done, with special attention to nanosilica and nanoclays. Regarding the effects of the modification with nanomaterials in the properties of the asphalt binder, few recent reviews address this topic. Porto et al. [5] presented a review about the asphalt binder modification covering several types of modifiers, such as, polymers, chemical modifiers (including nanocomposite modifiers), and warm mix technology. The review of Martinho and Farinha [51] focused on the use of nanoclays. Li et al. [52] presented a review covering a wide range of nanomaterials, such as, nanocarbon, nanoclay, nanofiber, nanosilica, and nanotitanium. Wu and Tahri [76] presented a state-of-the-art about the use of carbon and graphene family nanomaterials in asphalt modification.

The following paragraphs present a brief description of the effects of the modifications with nanomaterials in the properties of the asphalt binder, as well as, some additional details found in literature.

In brief, the nanosilica-modified binder presented a decrease in penetration, increase in viscosity, and increase in softening point [10,18,25,27,77,78]. Regarding the rheological behavior, evaluated using the dynamic shear rheometer (DSR), the modified binders present higher complex modulus and lower phase angle [18,64,77,79]. Authors evaluating the binder fatigue with DSR tests, concluded that the nanosilica modifications showed superior fatigue resistance [22,80,81].

At the level of fundamental characterization, the nanoclay-modified binder presented a decrease in penetration, increase in softening point, and increase in viscosity [10,25,27,35,39,41,55,79,82–89]. And consistently, regarding rheology, the nanoclay modified binders present an increase in complex shear modulus and decrease in phase angle [34,38,41,42,55,86,87,90]. The existing studies mostly focused on organically modified montmorillonite, due to the expectation of obtaining exfoliated structures in the modified binders and higher performance improvements. The raw nanoclays, in their hydrophilic natural form, may form only intercalated structures, although, some authors studying raw nanoclays [38,41,42,91,92] also obtained considerable performance improvements.

The type of nanoclay used in the modification has an important effect in results, i.e., in the modified binder performance. Although the overall trends were the same, the authors that studied more than one nanoclay type obtained different results, regardless of using the same control binder. A study [41] about the modification of 60/70 asphalt binder with nanoclays, sodium bentonite (BT) and organically modified sodium bentonite (OBT), revealed that both modifications caused reduction of the phase angle and increased viscosity, softening point, and complex shear modulus. The effects were correlated with the dosage of nanoclay introduced and, in all cases, the effects of the organically modified clay were stronger. The exfoliated structure of the organically modified nanoclay, that promotes a better dispersion in the asphalt matrix, can explain these effects. On the other hand, in another study [55], the authors studied the modification of a PG 64-28 with two organically modified nanoclays with similar structure (nanoclay A and nanoclay B) and observed different effects. For example, regarding viscosity, the modifications caused an increase of 41% and 112% with 2% of nanoclay A and 2% of nanoclay B, respectively, and regarding complex shear modulus, the modifications caused an increase of 66% and 184% with 2% of nanoclay A and 2% of nanoclay B, respectively. Jahromi and Khodaii [34] studied the effects of two organically modified nanoclays (Nanofil-15 and Cloisite-15A) on the properties of the 60/70 asphalt binder and, found the effects of the second stronger than those of the first. In the cases that authors studied the same dosage and nanomodification in different control binder, they reported that the effects in softer binders are stronger than those in harder binders [93,94].

Onochie et al. [79] studied the effects of the modifications with an organically modified nanoclay and nanosilica on the properties of a PG 58-28 asphalt binder. The results obtained by the authors were dependent of the type and dosage of the nanomaterial. For example, regarding viscosity, 2% nanoclay and 4% nanoclay increased the viscosity by 22% and 36%, respectively, and 2% nanosilica and 4% nanosilica increased the viscosity by 13% and 10%, respectively. Regarding complex shear modulus, 2% nanoclay and 4% nanoclay presented an increase of 19% and 40%, respectively, and 2% nanosilica and 4% nanosilica presented an increase of 21% and 35%, respectively.

A possible drawback of the modification of asphalt mixtures with nanomaterials is underperformance at low temperatures (PG low temperature). Onochie et al. [79] tested a PG 58-28 with 2% and 4% nanosilica modifications with bending beam rheometer (BBR) and found the modifications to present 6% and 14% higher creep stiffness and equal and 2% lower m-value, respectively. In the same study, the authors also evaluated the effects of 2% and 4% nanoclay modification, using the same control binder (PG 58-28) and an organically modified nanoclay (cloisite 30B), and found the modifications to present 8% and 14% higher creep stiffness and 2% and 4% lower m-value, respectively. Regarding nanoclays, other studies [38,55,90] reported similar underperformance of the modified binders in BBR or direct tensile test (DTT) at PG low temperature. However, in another study [41], the modifications of the 60/70 control binder with 5% nanoclay bentonite and 5% organically modified bentonite presented similar m-values but, 19% and 22% lower creep stiffness, respectively, indicating enhanced low temperature performance. On the one hand, the effects of the modifications with nanomaterials in the low temperature performance are not entirely understood and deserve further investigation. On the other hand, besides the worsening in low temperature performance, authors reported that the modified binders still passed the respective Superpave™ specification (maximum 300 MPa creep stiffness and minimum 0.300 m-value) presenting the same PG low temperature of the control binder. Thus, the use of such modifications in cold regions can be possible, but further investigation is recommended.

4. Effect of the Modification with Nanomaterials in the Mechanical Performance of Asphalt Mixtures

4.1. Nanosilica

The authors that conducted studies about nanosilica modified asphalt mixtures, reported important improvements in the mechanical performance of the mixture. To evaluate the asphalt mixture performance, the mechanical tests most found in literature were Marshall stability, water sensitivity (using the indirect tensile strength ratio or the retained Marshall stability), permanent deformation, and stiffness. The Figures 2–6 present the results found in several studies. "Control" refers to the mixtures with 0% of nanosilica and, "Modified" refers to the modification performed by each author.

Burguete [95] studied an asphalt concrete, AC 14 Surf 35/50, using limestone filler, basalt aggregate, and considering two binder contents, 5.5% and 6.5%, with three nanosilica modifications, 2%, 4%, and 6%, respectively. From Marshall tests results the author observed, for both binder contents, that the peak of Marshall stability corresponded to the 4% nanosilica modification and, across the dosage range, a decrease of the Marshall flow related with the increase of the nanosilica dosage. The results of Burguete [95] presented in Figure 2 correspond to the 6.5% binder content mixture with 4% nanosilica modification, that caused an increase of 19% in Marshall stability.

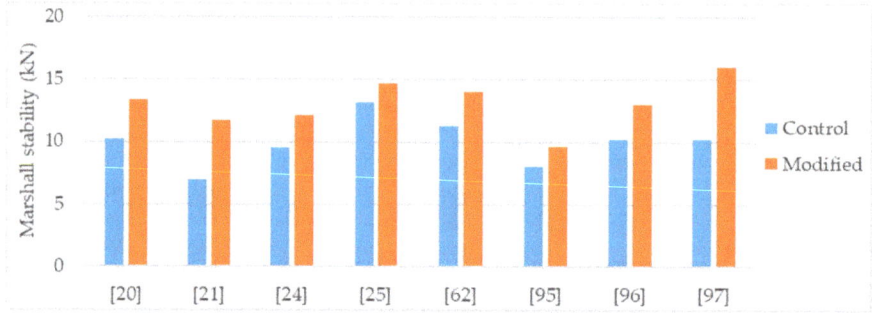

Figure 2. Marshall stability of control and nanosilica modified asphalt mixtures.

Sun et al. [96] performed Marshall tests in cores from two test tracks, (1) using as binder a neat bitumen AH-70 (penetration grade 60/70) modified with 5% of styrene-butadiene-styrene (SBS), and (2) using as binder the same neat bitumen (AH-70) modified with 5% styrene-butadiene-rubber (SBR), 0.5% nanosilica, and 1% polyethylene. The formula of the nanosilica-polymers modified mixture was determined through an orthogonal experiment. The authors concluded that a small amount of nanosilica could enhance significantly the performance of the polymer-modified binder. Regarding the Marshall stability results, with the nanosilica-polymers modification an increase of 28% was obtained (Figure 2).

Ghasemi et al. [21] studied a stone matrix asphalt (SMA), with aggregate maximum nominal size 19 mm and 6.3% binder content, modified with various percentages of nanosilica (0.5%, 1.0%, 1.5%, and 2.0%). The control binder used in the study was a 60/70 penetration grade bitumen modified with 5% SBS. The authors tested the asphalt mixtures for Marshall stability, indirect tensile strength, and indirect tensile stiffness modulus. The effects obtained in the tests increased according with the increase of nanosilica dosage. Thus, the mixture with 2% nanosilica had the most improved mechanical behavior. Regarding the 2% nanosilica modification, the Marshall stability increased 68% (Figure 2), indirect tensile strength increased 19% (Figure 3), and indirect tensile stiffness modulus increased 47%.

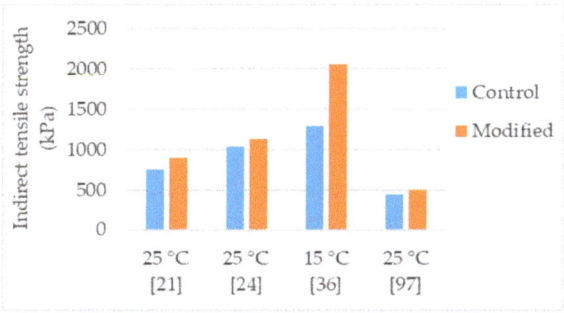

Figure 3. Indirect tensile strength of control and nanosilica modified asphalt mixtures.

Guo et al. [62] studied an asphalt mixture modified with 3% silane silica (nanosilica modified with silane coupling agent). The neat binder was the Panjin 90 (penetration grade 70/100) and alkaline aggregates were used for the mixtures production. The modified mixture presented 24% increase in Marshall stability (Figure 2) and 26% increase in dynamic stability number (rutting test). To address water sensitivity, the authors performed Marshall tests in specimens conditioned for 48 h immersion (IM) and for 48 h vacuum saturation (VS) and determined the respective retained Marshall stability. With both conditioning methods, the modified mixture performed better than the control (Figure 4).

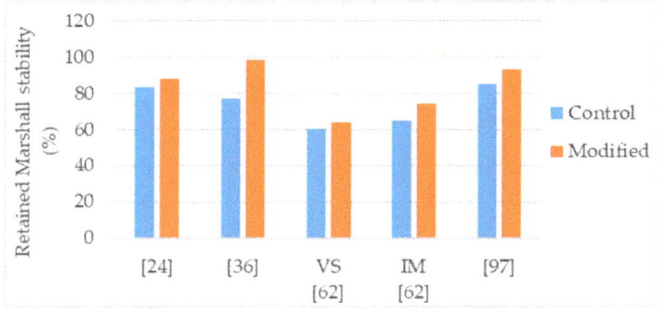

Figure 4. Retained Marshall stability of control and nanosilica modified asphalt mixtures.

Yusoff et al. [58] studied an asphalt mixture (NMAS 19 mm) with PG 76 polymer-modified binder and modified with 2% and 4% nanosilica. For the mixture production, granitic aggregates were used, and the selected binder content was 7.5%. The effects obtained with the modifications were sensitive to the nanosilica dosage. Thus, the 4% modification had the highest effects in the mechanical performance. The authors evaluated water sensitivity using the tensile strength ratio and concluded that the 4% nanosilica modification enhanced significantly the behavior of the mixture, raising TSR from 82% to 98% (Figure 5). Regarding resilient modulus, the 4% modification lead to 74% and 142% of increase (Figure 6), at the temperatures of 25 °C and 40 °C, respectively. Testing for dynamic creep, a 42% reduction in permanent deformation was observed.

Hasaninia and Haddadi [20] studied an asphalt mixture (NMAS 12.5 mm) modified with 2%, 4%, 6%, and 8% nanosilica. The control binder was the 60/70. The optimum binder content (OBC) was calculated using the Marshal methodology, and the results showed a decrease in OBC proportional to the increase in nanosilica dosage. The OBC values were 5.5%, 5.3%, 5.2%, 5.0%, and 4.9% for the 0% (control mixture), 2%, 4%, 6%, and 8% nanosilica modified, respectively. Regarding the performance tests, the effects obtained with the modifications were sensitive to the nanosilica dosage thus, the 8% nanosilica modification had the highest effects. Concerning the 8% modification, the modified mixture presented 31% increase in Marshall stability (Figure 2), 37% increase in resilient modulus (Figure 6),

30% increase in indirect tensile strength, decrease in indirect tensile strength ratio from 98% to 73% (Figure 5), 71% increase in flow number (lower permanent deformation), and better fatigue resistance (18% increase in the strain for 10^5 cycles). As the authors produced the modified mixtures using the respective OBC, the decrease in water sensitivity (indirect tensile strength ratio) can be partially explained by the higher air void content of the nanosilica-modified mixtures (4.47% air void content in control mixture versus 5.93% air void content in 8% nanosilica modification).

Yao et al. [22] studied an asphalt mixture with PG 58-34 with acrylonitrile-butadiene-styrene (control binder) and modified with 4% and 6% nanosilica. The effects obtained were dependent of the nanosilica dosage. Thus, the 6% modification had the highest effects in the mechanical performance. Regarding the 6% modification, the authors reported 30% increase in the dynamic modulus, 34% reduction in permanent deformation (evaluated with the asphalt pavement analyzer rutting test) and 27% increase in resilient modulus (Figure 6).

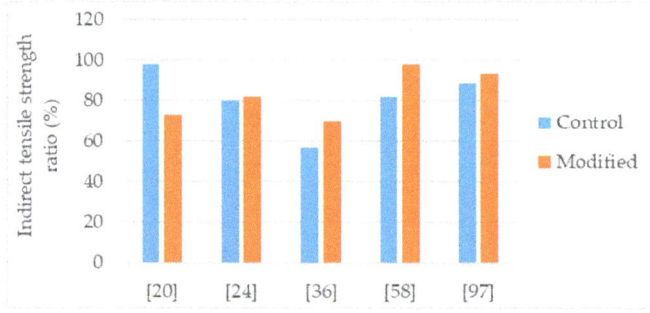

Figure 5. Indirect tensile strength ratio of control and nanosilica modified asphalt mixtures.

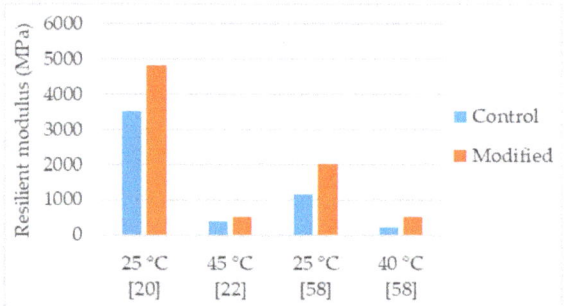

Figure 6. Resilient modulus of control and nanosilica modified asphalt mixtures.

Crucho et al. [25,36] studied an asphalt mixture (AC 14) with granitic aggregates, limestone filler, and 4.5% binder content. The selected binder was the 35/50 penetration grade (control binder) and the nanosilica dosage was 4%. The author concluded that the 4% nanosilica modification enhanced the mechanical performance of the mixture and reported the following effects: 11% increase in Marshall stability (Figure 2); improvement in retained Marshall stability from 77% to 98% (Figure 4); 59% increase in indirect tensile strength (Figure 3); improvement in indirect tensile strength ratio from 57% to 70% (Figure 5); better affinity aggregate-binder (more 10% of binder coverage); 34% reduction in permanent deformation; 4% and 18% average increase in stiffness modulus at the temperature of 20 °C and 30 °C, respectively; and better fatigue resistance (4% increase in the strain for 10^6 cycles).

Cai et al. [24] produced an asphalt mixture (AC 16) with limestone aggregates and 60/70 penetration grade bitumen and studied the modification with 1% nanosilica. The authors reported the following effects: 27% increase in Marshall stability (Figure 2); 14% reduction in permanent deformation;

improvement in retained Marshall stability from 83% to 88% (Figure 4); 9% increase in indirect tensile strength (Figure 3); small improvement in indirect tensile strength ratio from 80% to 82% (Figure 5); 14% increase in stiffness modulus and better fatigue life (97% increase in number of cycles) in fatigue tests at 1000 μm/m controlled strain; 9% increase in flexural tensile strength and 25% increase in flexural strain in the three-point bending test.

Ezzat et al. [97] studied an asphalt mixture (NMAS 19 mm) modified with 7% nanosilica. The asphalt mixture was produced with limestone aggregates, limestone filler, 60/70 (control binder) and 5.5% binder content. The nanosilica modified mixture presented 56% increase in Marshall stability (Figure 2), increase in retained Marshall stability from 85% for control to 93% for nanosilica modified (Figure 4), 11% increase in indirect tensile strength (Figure 3), and increase in indirect tensile strength ratio from 88% for control to 93% for nanosilica modified (Figure 5).

Tanzadeh and Shahrezagamasaei [23] studied a porous asphalt mixture (NMAS 12.5 mm). For the mixture production, the authors used limestone aggregates, limestone filler, several combinations of additives (glass fiber, lime powder, polypropylene fiber, SBS and nanosilica), 60/70 asphalt binder and the binder contents 4.5%, 5.5%, and 6.0%. For the tested additives combinations, the introduction of 2% and 4% nanosilica modifications caused reduction in binder drain down, reduction in permanent deformation, and increase in indirect tensile strength. The effects of the nanosilica increased according with the dosage increase. In a following study, Tanzadeh et al. [98] evaluated the performance of a porous asphalt mixture (NMAS 12.5 mm), produced with limestone aggregates, limestone filler, lime (0.5% by weight of mixture), and 60/70 asphalt binder modified with 4.5% SBS. The binder contents of the asphalt mixture were 5% and 6%. The authors study the modification with 2% nanosilica alongside with 0.2% glass fiber and basalt fiber. The introduction of 2% nanosilica increased indirect tensile strength, improved the resistance to moisture sensitivity, and reduced abrasion (Cantabro test).

The use of silica nanoparticles to modify the asphalt mixture revealed to cause an overall improvement in its mechanical performance. Generally, the studies found in literature indicate higher Marshall stability, higher indirect tensile strength, enhanced water sensitivity (indicated by higher indirect tensile strength ratio and/or higher retained Marshall stability), higher stiffness (higher resilient modulus or stiffness modulus), lower permanent deformation, and better fatigue resistance.

4.2. Nanoclay

In the studies of asphalt mixtures modified with nanoclays, several improvements in the mechanical performance were identified. Generally, the effect reported by the authors were: Increase in Marshall stability, reduction in permanent deformation, lower water sensitivity (increase in indirect tensile strength ratio and/or retained Marshall stability), increase in stiffness modulus/resilient modulus, and better resistance to fatigue. Although the findings are generally consistent, the big variety of materials leads to different effects in mechanical performance. The nanoclay type, the modification of raw nanoclay with organo-modifiers, the nanoclay dosage and the original asphalt binder properties can have a strong influence in the results. The following paragraphs describe the most relevant studies found in the literature and their more important conclusions.

Gedafa et al. [94] studied an asphalt mixture (NMAS 12.5 mm) with 5.7% binder content, using two different binders, PG 58-28 and PG 64-28 (control binders), and modified with an organically modified nanoclay (cloisite) with the dosages of 1%, 3%, and 5%. The authors tested for permanent deformation using the asphalt pavement analyzer rut test (APA), and concluded that the resistance to permanent deformation increased with the nanoclay dosage (Figure 7). The effect of the modifications in the softer binder (PG 58-28) was stronger that in the harder binder (PG 64-28). The modifications with 5% nanoclay presented a reduction of 57% and 38% with the PG 58-28 and PG 64-28, respectively.

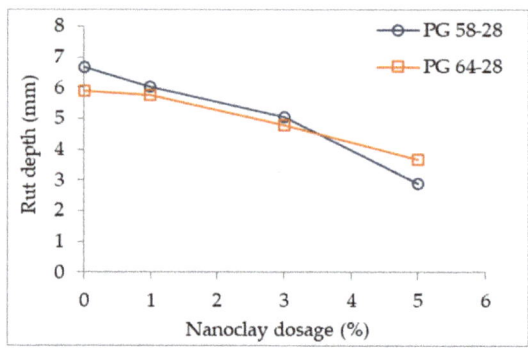

Figure 7. Effect of nanoclay dosage on rut depth [94].

Iskender [40] studied the effects of three nanoclay modifications in the performance of a SMA mixture. The SMA mixture (NMAS 12.5 mm) was produced with 6.1% binder content and, basalt aggregates and asphalt binder 50/70 (control binder) were used. The author introduced the nanoclay as a partial substituent of the mineral filler, and not by previously blending with the asphalt binder. In this case, the nanoclay dosages were 2%, 3.5%, and 5% by mass of dry aggregates. The author modified raw bentonite clay with three organic modifiers, obtaining three organically modified clays: Nanoclay A—modified with dimethyl, dehydrogenated tallow, quaternary ammonium (NC A); nanoclay B—modified with methyl, tallow, bis 2 hydroxyethyl quaternary ammonium (NC B); and nanoclay C—modified with dimethyl, benzyl, hydrogenated tallow, quaternary ammonium (NC C). The author evaluated water sensitivity using the modified Lottman test at 25 °C, and observed: With nanoclay A the indirect tensile strength increased by 14%, 20%, and 3%; with nanoclay B the indirect tensile strength increased by 11%, 20%, and −2%; and with nanoclay C the indirect tensile strength decreased by 11%, 14%, and 17%, for 2%, 3.5%, and 5% dosages, respectively. The indirect tensile strength results corresponding to the 2% modifications are presented in Figure 8. Regarding the indirect tensile strength ratio, for all the nanoclays, the highest ratio corresponded to the 2% dosage (Figure 9). The ratio presented a trend of decrease with the increase in dosage, in such way, with 5% dosage all nanoclays performed worse that the control mixture. The author evaluated permanent deformation at 40 °C using the dynamic creep test, applying 100 kPa load for 21,600 pulses. The nanoclay A enhanced the resistance to permanent deformation, presenting a reduction of 9%, 10%, and 36% for 2%, 3.5%, and 5% dosages, respectively. The mixture modified with nanoclay B performed worse that the control mixture, presenting an increase in permanent deformation of 30%, 40%, and 51% for 2%, 3.5%, and 5% dosages, respectively. The behavior of the mixture modified with nanoclay C was more complex, presenting 12% increase with 2% dosage, 29% reduction with 3.5% dosage, and similar to control with 5% dosage.

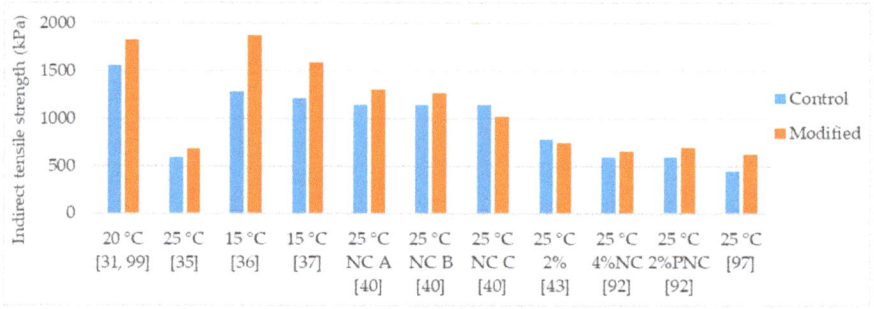

Figure 8. Indirect tensile strength of control and nanoclay modified asphalt mixtures.

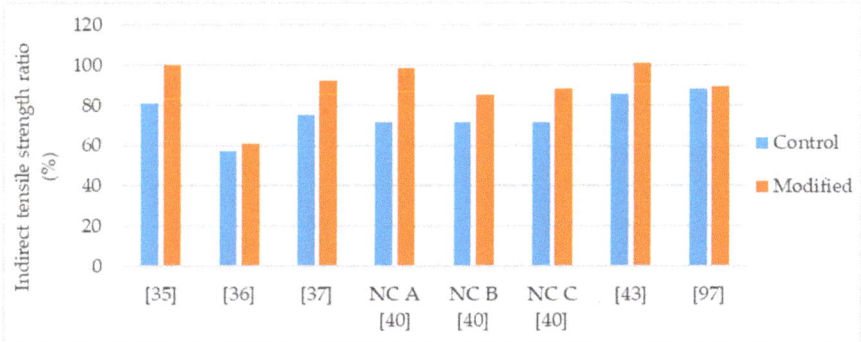

Figure 9. Indirect tensile strength ratio of control and nanoclay modified asphalt mixtures.

Ghile [31] and van de Ven et al. [99] studied an asphalt mixture (NMAS 11.2 mm) with 5.7% binder content, modified with of 6% organically modified nanoclay (cloisite). For the mixture production, granitic aggregates and 40/60 asphalt binder (control binder) were used. The modified mixture presented 12% average increase in resilient modulus in the range from 5 °C to 35 °C (evaluated by the indirect tensile resilient modulus test). Regarding indirect tensile strength, the modified mixture presented 8%, 19%, 17%, 28%, and 39% increase for 5 °C, 12.5 °C, 20 °C, 27.5 °C, and 35 °C, respectively (Figure 8). The modification also enhanced the resistance to permanent deformation (50% increase in the flow number in dynamic creep tests at 60 °C and 200 KPa).

Ezzat et al. [97] studied an asphalt mixture (NMAS 19 mm) modified with 3% nanoclay (montmorillonite). The asphalt mixture was produced with limestone aggregates, limestone filler, 60/70 (control binder) and 5.5% binder content. The authors observed that the nanoclay modified mixture presented 55% increase in Marshall stability (Figure 10), decrease in retained Marshall stability (85% for control and 77% for nanoclay modified), 40% increase in indirect tensile strength (Figure 8), and similar indirect tensile ratio (Figure 9).

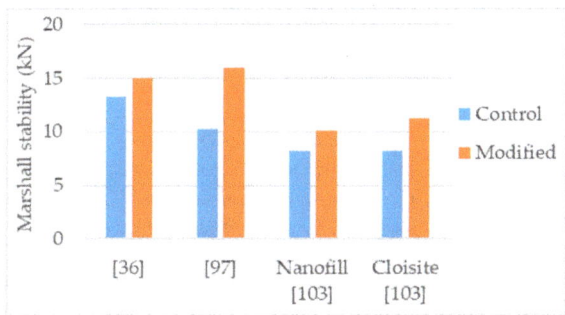

Figure 10. Marshall stability of control and nanoclay modified asphalt mixtures.

Crucho et al. [25,36,100] studied an asphalt mixture modified with 4% nanoclay hydrophilic bentonite. The mixture was an asphalt concrete (AC 14) using granitic aggregates, limestone filler, 35/50 asphalt binder, and 4.5% binder content. The modification had a positive effect in the properties of the asphalt mixture. The following effects were reported: 13% increase in Marshall stability (Figure 10); improvement in retained Marshall stability from 77% to 99%; 46% increase in indirect tensile strength (Figure 8); improvement in indirect tensile strength ratio from 57% to 61% (Figure 9); improvement in affinity aggregate-binder (from 40% to 67% of binder coverage); 10% reduction in permanent deformation (Figure 11); 7.2% reduction in stiffness modulus at 20 °C and 4.3% increase in stiffness modulus at 30 °C; and better fatigue resistance (7% increase in the strain for 10^6 cycles).

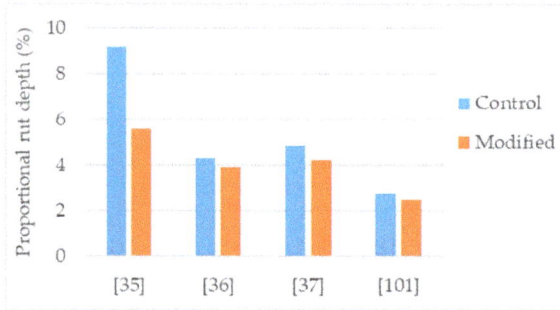

Figure 11. Proportional rut depth of control and nanoclay modified asphalt mixtures.

De Melo and Trichês [35] studied the modification of a PG 58-22 asphalt binder with 3% organically modified montmorillonite. The asphalt mixture, NMAS 19 mm, had 4.35% binder content and, was produced with basaltic aggregates, and hydrated lime. The modified mixture presented: 15% increase in indirect tensile strength (Figure 8), improvement in indirect tensile strength ratio from 81% to 100% (Figure 9); 39% reduction in permanent deformation (Figure 11); 20% and 57% average increase in stiffness modulus at the temperature of 20 °C and 30 °C, respectively; and increase in fatigue resistance (22% increase in the strain for 10^6 cycles).

Blom et al. [101] studied an asphalt mixture (AC14) modified with 5% organically modified nanoclay (cloisite-15). For the mixture production, sandstone aggregates and 35/50 asphalt binder were used. The authors tested for permanent deformation using a wheel-tracking machine. After applying 50,000 cycles, the 5% nanoclay modified mixture presented 10% reduction in permanent deformation (Figure 11).

Goh et al. [43] studied the effects of the modification with 1% and 2% polysiloxane-modified montmorillonite in the indirect tensile strength of an asphalt mixture. The mixtures were produced with 5.2% binder content and the PG 58-28 was the control binder. On the one hand, in the indirect tensile strength of the unconditioned (dry) specimens, the modified mixtures presented 1% and 4% reduction for the 1% and 2% dosage (Figure 8), respectively. On the other hand, in the indirect tensile strength of the conditioned (wet) specimens (standard conditioning in water according to AASHTO T283-03), the modified mixtures presented 7% and 13% increase for the 1% and 2% dosage, respectively. This led to a progressive increase of the indirect tensile strength ratio with the increase in nanoclay dosage, 86%, 92%, and 101% for the 0%, 1%, and 2% dosage (Figure 9), respectively. The authors found the asphalt mixtures modified with polysiloxane-modified montmorillonite less susceptible to deicing solutions. These findings are in good agreement with another study [102], that reported the nanoclay to have a positive effect in mitigating stripping damage of asphalt mixtures exposed to non-chloride deicer solutions.

Jahromi et al. [103] studied the modification of an asphalt mixture, NMAS 12.5 mm, with two types of organically modified montmorillonite nanoclay, nanofil, and cloisite, with the dosages of 2%, 4%, and 7%. Limestone aggregates, limestone filler, and 60/70 control binder were used in the production of the mixtures. The authors concluded that the effects of the modifications increased according to the increase in the nanoclay dosage and the effects of cloisite were stronger than those of nanofil. Regarding the modifications with 7% nanoclay, the Marshall stability increased by 23% and 37% for the nanofill and cloisite (Figure 10), respectively. The increase in the nanoclays dosage lead to an increase in the optimum binder content, possibly explained by the large surface area of the nanoclays. The 7% modified mixtures presented 8% and 6% increase in indirect tensile strength at 25 °C, 40% and 18% increase in resilient modulus at 40 °C, and reduction in permanent deformations (37% and 83% increase in the flow number in dynamic creep tests at 60 °C and 300 KPa), for cloisite and nanofill, respectively.

Golestani et al. [39] studied the effect of the modification with 1.5% organically modified montmorillonite (cloisite) in the properties of an asphalt mixture (NMAS 19 mm) with 4.8% binder content. For the production of the mixture the authors used limestone aggregates and considered two control binders, a neat PG 58-10 and a polymer modified binder (PG 58-10 modified with 6% SBS). Regarding the effects in the PG 58-10 binder (NMA), the modification lead to 10% increase in resilient modulus at 25 °C (Figure 12), 18% increase in resilient modulus at 40 °C (Figure 12), 26% reduction in permanent deformation, similar (0.5% increase) indirect tensile strength and improvement in indirect tensile strength ratio from 58% to 61%. Regarding the effects in the polymer modified binder (NCMA), the modification lead to 3% increase in resilient modulus at 25 °C (Figure 12), 12% increase in resilient modulus at 40 °C (Figure 12), 29% reduction in permanent deformation, similar (0.5% increase) indirect tensile strength, and improvement in indirect tensile strength ratio from 69% to 72%.

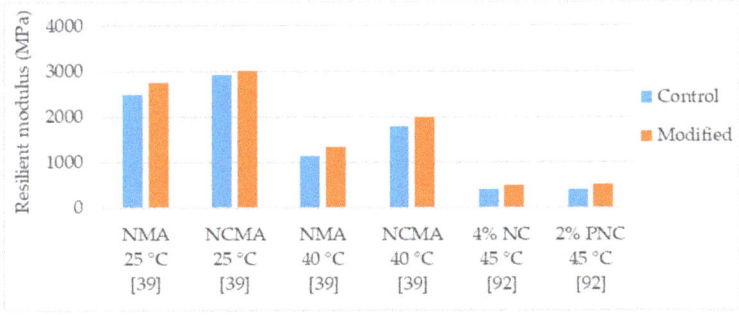

Figure 12. Resilient modulus of control and nanoclay modified asphalt mixtures.

Yao et al. [92] evaluated the effects of the modifications with non-modified montmorillonite nanoclay (NC) and polymer-modified montmorillonite (PNC) in the performance of an asphalt mixture (NMAS 9.5 mm). The control binder was the PG 58-34 and were studied the dosages of 2% and 4%. The modification with NC presented stronger effects with the higher dosage (4%) while the modification with PNC, with exception of permanent deformation test, presented higher effects with the 2% dosage. With the 4% NC modification, the authors reported 9% increase in indirect tensile strength (Figure 8), 15% average increase in dynamic modulus, 25% increase in resilient modulus (Figure 12) and 59% reduction in permanent deformation (rut depth using APA rutting test). Regarding the 2% PNC modification, the modified mixture presented 16% increase in indirect tensile strength (Figure 8), 10% average increase in dynamic modulus, 29% increase in resilient modulus (Figure 12) and 34% reduction in permanent deformation (rut depth using APA rutting test). Only in the case of permanent deformation, the 4% PNC performed better than the 2% PNC, presenting 42% reduction in rut depth.

Crucho et al. [37] studied a stone mastic asphalt (SMA16) mixture modified with 4% nanoclay hydrophilic bentonite. The mixtures were produced with 6.0% binder content and using 35/50 asphalt binder (control binder), granitic aggregates, and limestone filler. The mixture with 4% nanoclay presented: 30% increase in indirect tensile strength (Figure 8); improvement in indirect tensile strength ratio from 75% to 92% (Figure 9); 13% reduction in permanent deformation (Figure 11); 6.6% reduction in stiffness modulus at 20 °C; and better fatigue resistance (14% increase in the strain for 10^6 cycles).

Ameri et al. [104] modified a SMA mixture (NMAS 9.5 mm) with 1%, 2%, 3%, and 4% cloisite-15A. For the mixture production, 6.7% binder content, limestone aggregates and limestone filer were used. In this study, two control binders were considered, the 60/70 asphalt binder and the 60/70 modified with 5% SBS. In both mixtures, 60/70 and 60/70 + SBS, the nanoclay modification enhanced the properties of the mixture, increasing the effect with the increase in dosage. Thus, the 4% dosage presented the highest gains in performance. Regarding the mixture with 60/70 control binder, the 4% modification presented 23% increase in Marshall stability, 9% increase in flow number (lower plastic deformation) and 61% reduction in permanent deformation. Regarding the mixture with 60/70 + SBS control binder, the 4% modification presented 11% increase in Marshall stability, 8% increase in flow number (lower plastic deformation), and 16% reduction in permanent deformation.

Generally, the asphalt mixtures modified with nanoclays revealed higher Marshall stability, enhanced water sensitivity (indicated by higher indirect tensile strength ratio and higher retained Marshall stability), lower permanent deformation, and better fatigue resistance. Regarding stiffness, the nanoclay modified mixtures presented higher values, particularly at higher temperatures. The effects obtained with the modifications were dependent on the type of nanoclay being used. The raw nanoclays tend to present stronger effects according to the increase in dosage, while some of the organically modified nanoclays present a peak in performance enhancement at a relatively low dosage and a further increase worsens the performance. Although the majority of the nanoclay modifications had the effect of increasing the indirect tensile strength, some of organic modifiers caused a reduction.

4.3. Nanoiron

Kordi and Shafabakhsh [49] evaluated the mechanical properties of a stone mastic asphalt (SMA) mixture modified with 0.3%, 0.6%, 0.9%, and 1.2% of nano Fe_2O_3. The 60/70 asphalt binder was the control binder used in the study and, all the SMA were produced with 6.6% binder content. The authors tested for permanent deformation using the wheel-tracking test, for stiffness using the indirect tensile stiffness modulus (ITSM) and for fatigue using the indirect tensile fatigue test (ITFT). In all the tests, initially the performance improved, reaching peak with the 0.9% dosage, and then decreased for higher dosage. Regarding the 0.9% nano Fe_2O_3 modification; stiffness modulus increased by 28%, 14%, and 45%, at 5 °C, 15 °C, and 25 °C, respectively; fatigue life increased between 15% and 35%, depending on the temperature and stress level; and permanent deformation decreased by 48%, 35% and 18%, at 40 °C, 50 °C, and 60 °C, respectively.

Pirmohammad et al. [50] studied an asphalt mixture (NMAS 12.5 mm) modified with nano Fe_2O_3. The 60/70 asphalt binder (PG 64-22) was modified with the nanomaterial dosages of 0.1%, 0.4%, 0.8%, and 1.2%. The authors studied the fracture properties of the asphalt mixture by conducting semi-circular bending (SCB) tests. Regarding the nano Fe_2O_3 modification, the fracture resistance initially increases reaching the peak at 0.8% dosage (27% increase), and then decreases for higher dosage.

Crucho et al. [25,36] studied an asphalt concrete (AC 14) mixture modified with 4% nanoiron (zero-valent iron). For the mixture production was used granitic aggregates, limestone filler, 35/50 asphalt binder (control binder), and 4.5% mixture binder content. The modification had the following effects: 5% increase in Marshall stability; improvement in retained Marshall stability from 77% to 80%; 5% increase in indirect tensile strength; improvement in indirect tensile strength ratio from 57% to 82%; improvement in affinity aggregate-binder (from 40% to 70% of binder coverage); 20% reduction in permanent deformation; 7.3% and 1.5% reduction in stiffness modulus at 20 °C and 30 °C, respectively; and better fatigue resistance (4% increase in the strain for 10^6 cycles).

The authors studying the use of nanoiron to modify the asphalt mixture indicate several improvements in mechanical performance, such as, higher Marshall stability, higher indirect tensile strength, enhanced water sensitivity (indicated by higher indirect tensile strength ratio and higher retained Marshall stability), lower permanent deformation, better fatigue, and fracture resistance.

5. Aging Resistance

5.1. Asphalt Binder

The organic nature of the asphalt binder leads to a continuous non-reversible aging process. During the aging process, the chemical composition of the asphalt binder changes, generally, the asphaltenes content tend to increase, while resins and aromatics tend to decrease [1]. Regarding the properties of the binder, the aging process is most prominent in the form of hardening and brittleness. The aging of the asphalt mixtures is highly dependent on the aging of the asphalt binder thus, several methods were developed and standardized to simulate in laboratory the aging of the asphalt binder. Actually, the paving industry relies on such methods to establish criteria for the paving grade bitumen in Europe [105] and for the performance-graded asphalt binders in USA [106]. Also, due to its convenience, less time, and material consuming, researchers frequently only study aging at the binder level. The aging methods more frequently used are the rolling thin film oven test (RTFOT), thin film oven test (TFOT), and pressure aging vessel (PAV). The RTFOT is considered to be representative of the short-term aging (corresponding to the plant mixing and field compaction processes) and PAV is added to simulate long-term aging.

After the aging simulation, the properties of the aged binder can be compared with those of the unaged binder. Generally, researchers consider parameters such as retained penetration—RP (Equation (1)), increase in softening point—ISP (Equation (2)), viscosity aging index—VAI (Equation (3)), DSR rutting parameter ($G^*/\sin\delta$), and DSR fatigue parameter ($G^*\cdot\sin\delta$).

$$RP = \frac{\text{Aged penetration value}}{\text{Unaged penetration value}} \times 100 \qquad (1)$$

$$ISP = \text{Aged softening point value} - \text{Unaged softening point value} \qquad (2)$$

$$VAI = \frac{\text{Aged viscosity value} - \text{Unaged viscosity value}}{\text{Unaged viscosity value}} \times 100 \qquad (3)$$

Regarding the properties of the asphalt binder, the aging process causes an increase in viscosity and softening point and a reduction in penetration value. From the rheology perspective, the aging process causes an increase in the complex modulus and a decrease in the phase angle.

Several authors highlight the positive contributions of nanomaterials to aging resistance. The presence of nanomaterials may play an important role in improving the asphalt mixture aging resistance by two mechanisms, first they may act as a barrier thus retarding the oxidation process and second as they can prevent the evaporation of light components of the asphalt [53].

Jahromi and Khodaii [34] studied the 60/70 asphalt binder modification using two types of organically modified nanoclays, cloisite, and nanofil, with the dosages of 2%, 4%, and 7%. The modifications lead to higher RP and lower ISP values, indicating less effect of aging in the binder. The enhancement in aging resistance was increasing according to the increase in nanoclay dosage (Figure 13). Regardless of both nanoclays were organically modified, the obtained effects were dependent of the nanoclay type. According with both parameters, RP and ISP, the modifications with nanofil were more effective in enhancing aging resistance. Other studies concerning nanoclay modifications, reported similar trends, increase in RP and reduction in ISP and VAI [79,84,107–110].

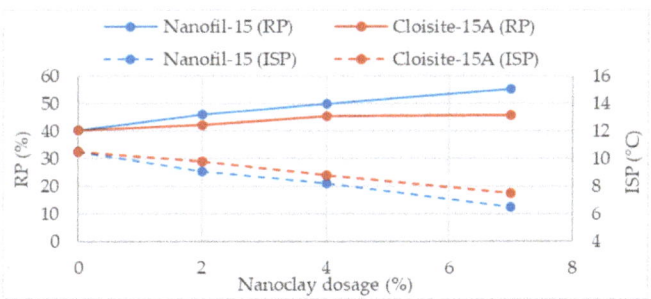

Figure 13. Retained penetration and softening point increment of two nanomodified binders [34].

Ghile [31] and van de Ven et al. [99] studied the effects of organically modified nanoclays in the rheology of two asphalt binders. The authors selected 6% nanofill to modify the 70/100 asphalt binder and, 3% and 6% cloisite to modify the 40/60 asphalt binder. After short-term and long-term aging, in DSR testing the modified binders showed smaller variations (smaller increase in stiffness and smaller reduction of phase angle) when compared with the unaged modified binder, which means less aging effect. Then, the variation of the rutting parameter (G*/sinδ) was also smaller in the case of the modified binders. The modified binders also showed higher retained penetration and lower increment in softening point. Other studies [111,112] evaluating aging through the variation of rheological parameters, obtained similar conclusions. Ashish, Singh and Bohm [113] modified a 35/50 asphalt binder with 2%, 4%, and 6% of cloisite-30B (organically modified nanoclay). The authors evaluated fatigue of the PAV aged binders using the linear amplitude sweep (LAS) test and found the number of load cycles to failure increasing with the increasing of nanoclay dosage. Similar results were presented by Kavussi and Barghabany [114], that modified a PG 58-22 with 2% and 6% of cloisite-30B.

In addition to the traditional methods, that essentially reproduce well the aging caused by temperature and oxidation mechanisms, several researchers are developing alternative laboratory aging simulations to address other effects. Environmental conditions, such as, solar radiation, particularly the ultraviolet (UV), and moisture damage also have a significant role in the change of binder morphology during the aging process [115–128].

Zhang et al. [110] concluded that the modification of an 60/80 asphalt binder with 6% organically modified montmorillonite enhanced the resistance to UV aging. After UV aging, the modified binder presented lower VAI and SPI than the control binder. Other studies presented similar conclusions [129,130].

Crucho et al. [131] tested the recovered binders of TEAGE aged (under UV radiation and watering/drying cycles) [132] asphalt mixture specimens. The mixtures were modified with three nanomaterials independently, 4% nanosilica, 4% nanoiron, and 4% nanoclay bentonite (hydrophilic clay). The results obtained for RP were 55%, 76%, 55%, and 54% for the control, 4% nanosilica, 4% nanoiron, and 4% nanoclay, respectively. The results obtained for ISP were 14.5 °C, 12.2 °C, 14.3 °C, and 13.5 °C for the control, 4% nanosilica, 4% nanoiron, and 4% nanoclay, respectively. All the modifications presented lower ISP, although, the results of nanoiron modification are similar to those of control, indicating no enhancement in aging resistance. Regarding RP, the control, 4% nanoiron, and 4% nanoclay also presented very similar results. Other the other hand, 4% nanosilica modification presented a clear indication of enhanced aging resistance according both parameters.

5.2. Asphalt Mixtures

5.2.1. Methods

The evaluation of the asphalt mixture aging is complex and, currently, there is no test method to assess it directly. A possible approach is to characterize the asphalt mixture's mechanical performance at early age (fresh mixture) and under aged conditions and then, determine the difference between them. Alternatively, some researchers compute the percent variation (principle similar to the VAI, Equation (3)) or the ratio between aged and fresh (principle similar to a retained performance, Equation (1)), commonly designating them as aging sensitivity and aging index, respectively. In this review, the effects of the nanomaterials in the aging resistance were analyzed by calculating the aging sensitivity (Equation (4)) of the evaluated mechanical performance parameters.

$$\text{Aging sensitivity} = \frac{\text{Aged value} - \text{Unaged value}}{\text{Unaged value}} \times 100 \tag{4}$$

Apparently, the higher the absolute value of the selected indicator (difference aged-unaged, aging sensitivity, or aging index) the most aging sensitive is the mixture. A drawback of this approach is that in the case of a mechanical property which value increases with aging, e.g. stiffness or tensile strength, in some cases, a poor aging resistance performance may be wrongly understood as a good aging resistance. As example, Figure 14 presents the case of a mixture (mixture B) with higher aging sensitivity than the control mixture (mixture A) but presenting lower aging index in all evaluations after age one. Thus, to draw conclusions about aging resistance, one should have a good overview of the performance under aged conditions by conducting several performance tests. In this regard, due to the brittleness effect caused by aging, testing for fatigue resistance is particularly important. The production of aged specimens consumes a significant amount of time and resources, thus, generally, it is more practical to perform different tests than to simulate several aging times.

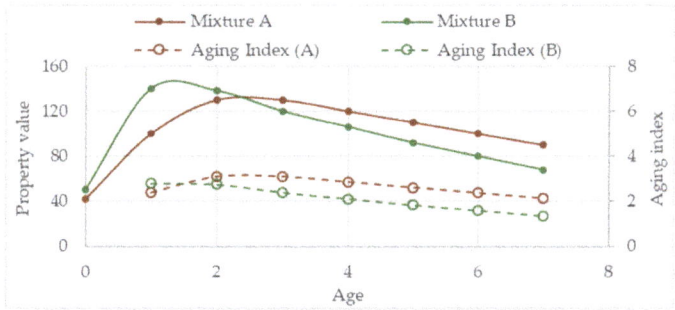

Figure 14. Example of the evolution of aging index.

To obtain aged asphalt mixture specimens few laboratory aging methods were developed. Currently, the methods used more often worldwide by the researchers are the described by the specification AASHTO R30-02 [133], initially proposed by the Strategic Highway Research Program [134]. In brief, the specification presents two methods, the short-term oven aging (STOA) and the long-term oven aging (LTOA). The STOA consists in conditioning the loose asphalt mixture in a draft oven for four hours at 135 °C and, aims to simulate the aging caused the plant mixing and compaction processes. The LTOA consists in conditioning the compacted asphalt specimens for five days at 85 °C and, aims to reproduce the aging of the pavement in service for approximately five to ten years. Although STOA and LTOA are practical methods requiring only simple equipment, they do not simulate all the actions present during field aging, e.g. solar radiation and moisture damage. Thus, researchers proposed few alternative approaches to address aging simulation, such as the saturation aging tensile stiffness (SATS) developed by Choi [135], the Viennese Aging Procedure (VAPro) developed by Steiner et al. [136], and the Tecnico accelerated aging (TEAGE) developed by Crucho et al. [132].

5.2.2. Nanosilica

Yusoff et al. [58] aged the asphalt mixture using STOA and LTOA methods and, tested aged specimens for resilient modulus (RM) at 25 °C and 40 °C, and permanent deformation using the dynamic creep test. The authors found that the 4% nanosilica modification provided some aging resistance particularly at the long-term aging. Figure 15 presents the aging sensitivity of the parameters considered by the authors and under both aging methods.

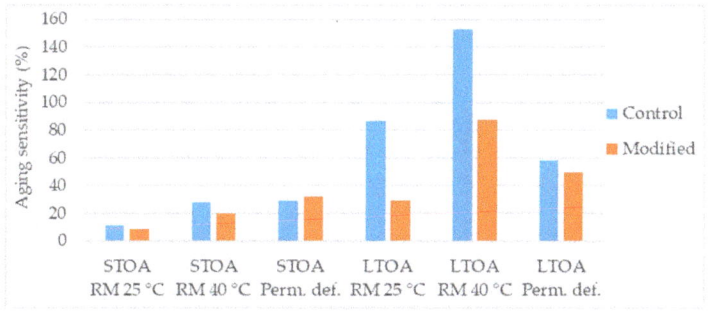

Figure 15. Aging sensitivity of the parameters considered in the study [58].

Cai et al. [24] produced LTOA aged specimens and conducted flexural tensile strength tests (three-point bending at −10 °C) and fatigue tests (four-point bending at 1000 μm/m constant strain). To study the effects of aging the authors compared the results of flexural strength, flexural strain, and fatigue life of the unaged and LTOA aged specimens and, concluded that the 1% nanosilica modification had a positive effect in aging resistance. Figure 16 presents the aging sensitivity of the parameters considered.

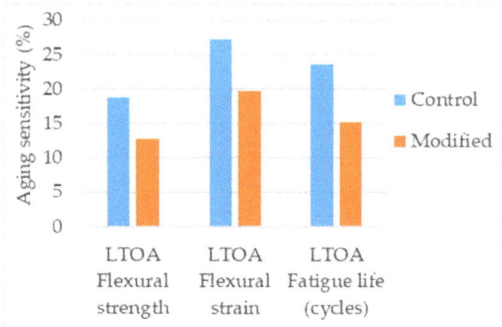

Figure 16. Aging sensitivity of the parameters considered in the study [24].

Crucho et al. [131] studied the effects of aging in the asphalt mixture using the TEAGE method [132], tuned to simulate seven years of field aging due to environmental conditions (UV radiation and precipitation) in the region of Lisbon. The aged mixtures were tested for indirect tensile strength (ITS), stiffness modulus, and fatigue, and the results were compared with those obtained under unaged conditions. The 4% nanosilica modified mixture presented lower aging sensitivity in all parameters, indicating an enhanced aging resistance (Figure 17).

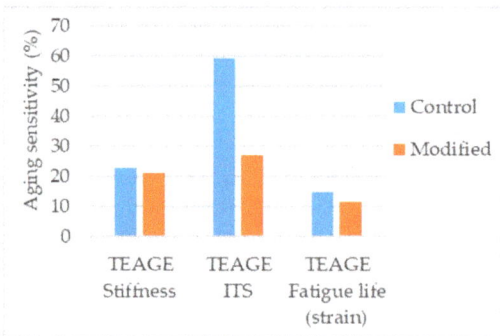

Figure 17. Aging sensitivity of the parameters considered in the study [131].

Guo et al. [62] produced STOA aged specimens and compared their performance to that of unaged specimens. The authors evaluated the retained Marshall stability (RMS), using two conditioning methods: (1) 48 h immersion (IM) and (2) 48 h vacuum saturation (VS), and the tensile strength ratio (TSR). Analyzing the aging sensitivity of these parameters (Figure 18) it can be concluded that the modified mixture (3% silane silica) suffered less variation in these ratios. This indication is positive, although would be not sufficient to property conclude about aging resistance, as it would be possible that both specimens conditioned/unconditioned suffer intense aging thus maintaining approximately similar ratio under aged and unaged conditions. The analysis of Marshall stability or indirect tensile strength would be preferable.

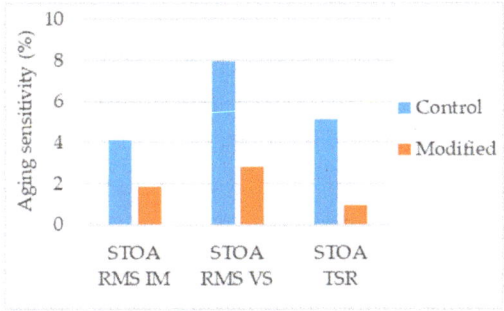

Figure 18. Aging sensitivity of the parameters considered in the study [62].

Tanzadeh et al. [98] evaluated the abrasion of a porous asphalt mixture (NMAS 12.5 mm), under aged and unaged conditions, using the Cantabro test. The asphalt mixtures were produced with limestone aggregates, limestone filler, lime (0.5% by weight of mixture), and 60/70 asphalt binder modified with 4.5% SBS. The binder contents of the asphalt mixture were 5% and 6%. The authors used 0.2% glass fiber and 0.2 basalt fiber to modify the asphalt mixture and, following, studied the modification with 2% nanosilica. The specimens were conditioned for aging in a draft oven at 60 °C for seven days. In the mixture with 0.2% glass fiber, the modification with 2% nanosilica presented a reduction in Cantabro loss from 10.1% (control) to 4.4% (modified) and, from 7.7% (control) to 3.5% (modified), for binder content of 5% and 6%, respectively. In the mixture with 0.2% basalt fiber, the modification with 2% nanosilica presented a reduction in Cantabro loss from 10.7% to 5.5% and, from 8.9% to 4.7%, for binder content of 5% and 6%, respectively.

Under a variety of mechanical tests and aging methods, the asphalt mixtures modified with nanosilica presented systematically lower aging sensitivity if compared to the correspondent control mixture, indicating enhanced aging resistance.

5.2.3. Nanoclay

López-Montero et al. [100] studied the effects of aging in the indirect tensile strength (ITS), in dry and wet (water sensitivity conditioned) conditions, of the asphalt mixture using the LTOA and TEAGE methods. The TEAGE method [132] was used to simulate seven years of field aging in the region of Lisbon. The authors determined an aging index, ratio between aged, and unaged values, and concluded that for both conditions, dry and wet, and under both aging methods, LTOA and TEAGE, the modification with 4% nanoclay bentonite presented less aging (Figure 19).

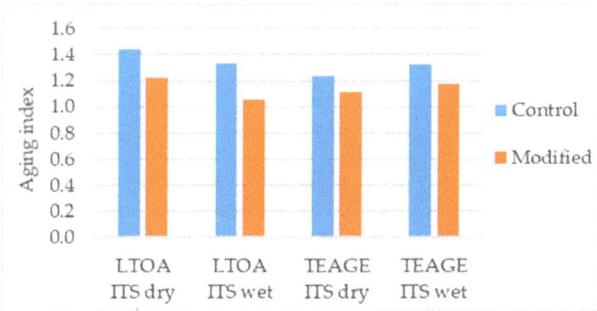

Figure 19. Aging index of the parameters considered in the study [100].

Crucho et al. [131] studied the effects of aging in an asphalt mixture modified with 4% nanoclay bentonite, using the TEAGE method, tuned to simulate seven years of field aging in the region of Lisbon.

The aged mixtures were tested for indirect tensile strength, stiffness modulus and fatigue. The modified mixture presented lower aging sensitivity in indirect tensile strength and fatigue resistance, and higher in the case of stiffness modulus. Nevertheless, as fatigue resistance is the most critical parameter under aged conditions, the results are a positive indication of enhanced aging resistance (Figure 20).

The number of studies about the effects of nanoclay modification in the aging of asphalt mixtures is still limited. Nevertheless, the results found in literature gave a positive indication of enhanced aging resistance, as in most cases, the asphalt mixtures modified with nanoclay presented lower aging sensitivity if compared to the correspondent control mixture.

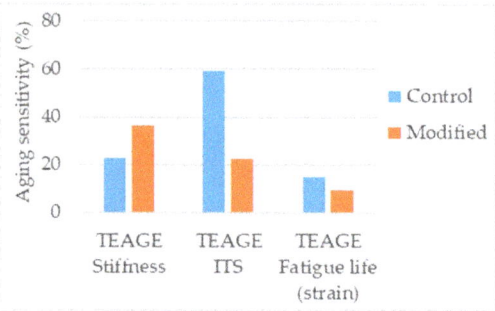

Figure 20. Aging sensitivity of the parameters considered in the study [131].

5.2.4. Nanoiron

Crucho et al. [131] studied the effects of aging in an asphalt concrete mixture modified with 4% nanoiron. The TEAGE [132] method was used to simulate seven years of field aging in the region of Lisbon. The aged mixtures were tested for stiffness modulus, indirect tensile strength, and fatigue resistance. If compared with the control asphalt mixture the nanoiron-modified mixture presented higher aging sensitivity in all the parameters considered (Figure 21). Nevertheless, it is not possible to conclude immediately that nanoiron has a negative effect in the aging resistance of the asphalt mixture. As this is the only study found in literature addressing nanoiron-modified aged asphalt mixture, more studies are needed to validate these findings. Other studies [49,50] that determined the optimum nanoiron (Fe_2O_3) dosage using mechanical performance tests, indicated the optimum dosage to be 0.8% to 0.9% and, further increase in dosage worsened the performance. Although in the study of Crucho et al. [131] the used nanoparticles were different (zero-valent iron), this is a strong indication that the dosage considered (4%) is too high, beyond the optimum.

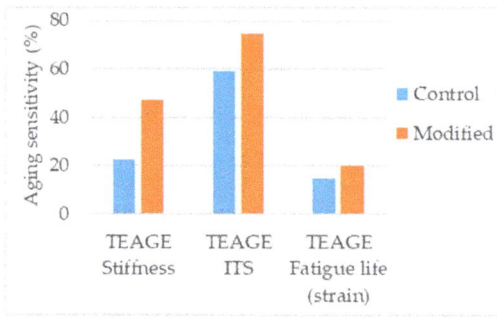

Figure 21. Aging sensitivity of the parameters considered in the study [131].

6. Cost Evaluation

Asphalt pavement construction deals with large quantities of materials, in the range of tons, contrasting with the nanotechnology that synthetizes and manipulates few grams of a specific material. As the production of nanomaterials is oriented to deliver small quantities of highly controlled materials, i.e., purity levels ≥98%, the cost of these materials is relatively high compared to traditional paving materials. The costs are justified by the use of expensive equipment and technology involved, as well as, the fact that nanomaterials production is generally conducted by highly qualified personal. However, this cost has been decreasing over time and, the trend is for further decrease, pushed by an increasing demand and improvements in manufacturing technology.

Generally, due to the costs associated with their synthesis, the price of the nanomaterials are highly dependent on the particle's size range and specific surface area. Products with high purity and narrow size range and high specific surface may demand higher processing efforts, thus having higher final costs.

The prices for nanosilica particles can vary from 80 €/kg to 1500 €/kg. The products with smaller size and narrow range (5 nm to 20 nm) and high specific surface (up to 690 m^2/g) can cost between 1000 €/kg and 1500 €/kg. On the other side, particles with a wider size range (15 nm to 70 nm) and higher variation on specific surface area (130 m^2/g to 600 m^2/g) can cost between 80 €/kg to 150 €/kg. Further modifications of the nanosilica, for example with silane coupling agents, also brings an increase to the costs. Silane modified silica nanoparticles can cost between 180 €/kg to 450 €/kg.

Currently, the commercial price of nanoclays ranges from 100 €/kg to 250 €/kg, depending on the particle size and types of treatment/modification. Organically modified nanoclays are more expensive due to the additional surface treatment process. On the other side, raw nanoclays with natural hydrophilic behavior are cheaper.

The prices of nanoiron particles are in the range of 100 €/kg to 2600 €/kg. Costs are dependent on purity level and size range. Nanoparticles of Fe_2O_3 with average size 50 nm and purity around 98% cost can cost between 100 €/kg to 300 €/kg. The most expensive iron nanoparticles are 99.9% 5 nm Fe_2O_3. The zero-valent nanoiron (50 nm) can cost nearly 120 €/kg.

The framework provided by Martinho et al. [137] was considered to perform a simple cost analysis regarding the construction of a 5 cm thick asphalt concrete (AC14 surf 35/50) wearing course. Using conventional materials (unmodified asphalt binder) the estimated cost was 6.8 €/m^2. Regarding the nanomodified binders, the cost where estimated using the costs of the respective raw materials but not accounting with the possible costs of technology implementation for industrial scale refinery- or plan-modification. Considering the available nanosilica with the lower cost, the asphalt mixture cost was estimated to be 21.0 €/m^2, 44.7 €/m^2 and 38.9 €/m^2 for the modifications with 3% nanosilica, 8% nanosilica, and 3% silane-nanosilica, respectively. The cost for nanoclay modifications was estimated to be 25.4 €/m^2 and 51.4 €/m^2, when using 3% non-modified nanoclay and 3% organically modified nanoclay (modified with quaternary ammonium), respectively. The cost for nanoiron modifications were estimated in 12.1 €/m^2 and 35.3 €/m^2 for the modifications with 0.9% nano-Fe_2O_3 and 4% nano zero-valent iron.

The low dosage required by the nano-Fe_2O_3 modification indicated that it is competitive from the economic point of view. Regarding the group of the most studied nanomaterials, nanosilica, and nanoclays, the cost analysis suggests that nanosilica may be more promising as it had the lowest cost. Nevertheless, the three-fold increase in cost might be justified by the improvements obtained in mechanical performance and aging resistance. At current prices, for the nanomodifications to be cost effective, a significant improvement in durability has to be attained. It is expected that the production of nanoparticles in bulk quantity and the use of alternative sources will significantly reduce the costs. For example, rice husk, an abundant waste biomass with high content of silica was identified as a potential low-cost resource for the production of nanosilica particles [138].

7. Concluding Remarks

This review addressed the effects of the modifications with nanomaterials: Nanosilica, nanoclay, and nanoiron, on the mechanical properties of the asphalt mixtures. The studies about modification of asphalt binders with nanomaterials showed that if a correct dispersion of the nanomaterials in the asphalt matrix is attained, several properties can be improved. Regarding the properties of the asphalt binders, the modifications with nanomaterials brought a reduction in penetration value and an increase of viscosity and softening point, as well as, increase in complex modulus and a decrease in phase angle. The effects of the modifications are dependent on the type and dosage of the nanomaterial, as well as, on the properties of the original asphalt binder to be modified. The optimum nanomaterial dosage may be dependent on the mechanical property to be enhanced, as well as the cost of the modification. For example, softer binders attained higher gains in resistance to permanent deformation with the nanomaterials modification than harder binders.

The effects of the nanomaterial modifications in the properties of the asphalt mixtures conducted to similar trends than for binders, meaning that the consideration of the type of modification is connected to the goal in terms of performance achieved for the best cost. The review carried out allows for the following conclusions:

- For asphalt mixtures, the modifications with nanosilica present higher Marshall stability, higher indirect tensile strength, higher stiffness modulus, lower permanent deformation, enhanced fatigue resistance, and higher resistance to moisture damage. The studies found in literature covered the dosage range of 0.5% to 8%, and the effects of the modifications increase according with the increase in dosage. Nevertheless, the cost of the modification should be taken into account, as the use of high dosages can return not economical.
- The effects of the modifications with nanoclays were dependent of the type of nanoclay. Besides the particle size distribution and the specific surface area, the type of treatment, raw (hydrophilic) or organically modified (hydrophobic), can have particular importance in the obtained effects. Generally, the effects of nanoclays modifications are higher Marshall stability, lower permanent deformation, higher stiffness modulus at high temperatures, better fatigue resistance, and higher resistance to moisture damage. The studies found in literature covered the dosage range of 1% to 7% and, the effects of the modifications increase according with the increase in dosage. Nevertheless, in the case of some organically modified nanoclays, the performance enhancement peaked at relatively low dosages (e.g. 2%) and a further increase in dosage worsened the performance. As the cost of organically modified nanoclays can be roughly double the cost of raw nanoclay, the use of such modifications should be careful optimized to ensure maximum performance with lowest dosage possible.
- Currently, the modifications with nanoiron are not as explored as those with nanosilica and nanoclays, nevertheless, several studies reported important improvements in the mechanical performance of the modified asphalt mixture, such as, higher Marshall stability, higher indirect tensile strength, higher resistance to moisture damage, lower permanent deformation, enhanced fatigue resistance, and fracture resistance. The optimum dosage of nanoiron particles seems to be in the range of 0.8% to 0.9%, where the performance gains peak and, further increase in dosage worsens the performance. Comparatively to other nanomaterials, the lower optimum dosage of nanoiron particles can make them more competitive from economic point of view.
- Besides the improvements in mechanical performance, the nanomaterials also gave good indications regarding an enhanced aging resistance. The study of the properties of aged asphalt mixtures highlighted the full potential of the nanomaterials and this is a good inference regarding the necessary life cycle cost evaluation in order to achieve a good balance between direct construction costs, high for now due to the use of nanomaterials, and the long term effect on durability.

Author Contributions: Conceptualization, J.C. and L.P.-S.; methodology, J.C., L.P.-S., J.N. and S.C.; validation, J.C. and L.P.-S.; formal analysis, J.N. and S.C.; investigation, J.C., J.N. and S.C.; resources, L.P.-S.; writing—original draft preparation, J.C.; writing—review and editing, J.C. and L.P.-S.; supervision, L.P.-S., J.N. and S.C.

Funding: This research received no external funding.

Conflicts of Interest: The authors declare no conflict of interest.

References

1. Hunter, R.N.; Self, A.; Read, J. *The Shell Bitumen Handbook*, 6th ed.; Shell Bitumen by ICE Publishing: London, UK, 2015.
2. Carraher, C.E., Jr. *Seymour/Carraher's Polymer Chemistry*, 6th ed.; Marcel Dekker: New York, NY, USA, 2003.
3. Erkens, S.; Porot, L.; Glaser, R.; Glover, C.J. Aging of Bitumen and Asphalt Concrete: Comparing State of the Practice and Ongoing Developments in the United States and Europe. In Proceedings of the Transportation Research Board 95th Annual Meeting, Washington, DC, USA, 10–14 January 2016.
4. Apostolidis, P.; Liu, X.; Kasbergen, C.; Scarpas, T.A. Synthesis of Asphalt Binder Aging and the State-of-the-Art of Anti-aging Technologies. *Transp. Res. Rec.* **2017**, *2633*. [CrossRef]
5. Porto, M.; Caputo, P.; Loise, V.; Eskandarsefat, S.; Teltayev, B.; Oliviero Rossi, C. Bitumen and Bitumen Modification: A Review on Latest Advances. *Appl. Sci.* **2019**, *9*, 742. [CrossRef]
6. The European Commission. EC Commission Recommendation of 18 October 2011 on the definition of nanomaterial (2011/696/EU). *Off. J. Eur. Union* **2011**, *L 275*, 38–40.
7. ASTM International. *ASTM E2456-06. Standard Terminology Relating to Nanotechnology*; ASTM: West Conshohocken, PA, USA, 2012.
8. Donegá, C.D.M. (Ed.) The Nanoscience Paradigm: "Size Matters!". In *Nanoparticles: Workhorses of Nanoscience*; Spinger: Berlin, Germany, 2014; pp. 1–12.
9. Steyn, W.J. Applications of Nanotechnology in Road Pavement Engineering. In *Nanotechnology in Civil Infrastructure*; Gopalakrishnan, K., Birgisson, B., Taylor, P., Attoh-Okine, N.O., Eds.; Springer-Verlag: Berlin/Heidelberg, Germany, 2011; pp. 48–83.
10. Ezzat, H.; El-Badawy, S.; Gabr, A.; Zaki, E.-S. Evaluation of asphalt enhanced with locally made nanomaterials. *Nanotechnologies Constr.* **2016**, *8*, 42–67. [CrossRef]
11. Yuvakkumar, R.; Elango, V.; Rajendran, V.; Kannan, N. High-purity nano silica powder from rice husk using a simple chemical method. *J. Exp. Nanosci.* **2014**, *9*, 272–281. [CrossRef]
12. Liou, T.H.; Yang, C.C. Synthesis and surface characteristics of nanosilica produced from alkali-extracted rice husk ash. *Mater. Sci. Eng. B Solid-State Mater. Adv. Technol.* **2011**, *176*, 521–529. [CrossRef]
13. Santana Costa, J.A.; Paranhos, C.M. Systematic evaluation of amorphous silica production from rice husk ashes. *J. Clean. Prod.* **2018**, *192*, 688–697. [CrossRef]
14. Ma, X.; Zhou, B.; Gao, W.; Qu, Y.; Wang, L.; Wang, Z.; Zhu, Y. A recyclable method for production of pure silica from rice hull ash. *Powder Technol.* **2012**, *217*, 497–501. [CrossRef]
15. Phoohinkong, W.; Kitthawee, U. Low-Cost and Fast Production of Nano-Silica from Rice Husk Ash. *Adv. Mater. Res.* **2014**, *979*, 216–219. [CrossRef]
16. Moosa, A.A.; Saddam, B.F. Synthesis and Characterization of Nanosilica from Rice Husk with Applications to Polymer Composites. *Am. J. Mater. Sci.* **2017**, *7*, 223–231.
17. Karnati, S.R.; Oldham, D.; Fini, E.H.; Zhang, L. Surface functionalization of silica nanoparticles to enhance aging resistance of asphalt binder. *Constr. Build. Mater.* **2019**, *211*, 1065–1072. [CrossRef]
18. Shafabakhsh, G.H.; Ani, O.J. Experimental investigation of effect of Nano TiO2/SiO2 modified bitumen on the rutting and fatigue performance of asphalt mixtures containing steel slag aggregates. *Constr. Build. Mater.* **2015**, *98*, 692–702. [CrossRef]
19. Hasaninia, M.; Haddadi, F. Studying Engineering Characteristics of Asphalt Binder and Mixture Modified by Nanosilica and Estimating Their Correlations. *Adv. Mater. Sci. Eng.* **2018**, *2018*. [CrossRef]
20. Hasaninia, M.; Haddadi, F. The characteristics of hot mixed asphalt modified by nanosilica. *Pet. Sci. Technol.* **2017**, *35*, 351–359. [CrossRef]
21. Ghasemi, M.; Marandi, S.M.; Tahmoorsi, M.; Kamali, R.J.; Taherzade, R. Modification of Stone Matrix Asphalt with Nano-SiO$_2$. *J. Basic Appl. Sci. Res.* **2012**, *2*, 1338–1344.

22. Yao, H.; You, Z.; Li, L.; Lee, C.H.; Wingard, D.; Yap, Y.K.; Shi, X.; Goh, S.W. Rheological Properties and Chemical Bonding of Asphalt Modified with Nanosilica. *J. Mater. Civ. Eng.* **2013**, *25*, 1619–1630. [CrossRef]
23. Tanzadeh, J.; Shahrezagamasaei, R. Laboratory Assessment of Hybrid Fiber and Nano-silica on Reinforced Porous Asphalt Mixtures. *Constr. Build. Mater.* **2017**, *144*, 260–270. [CrossRef]
24. Cai, L.; Shi, X.; Xue, J. Laboratory evaluation of composed modified asphalt binder and mixture containing nano-silica/rock asphalt/SBS. *Constr. Build. Mater.* **2018**, *172*, 204–211. [CrossRef]
25. Crucho, J.; Neves, J.; Capitão, S.; Picado-Santos, L. Mechanical performance of asphalt concrete modified with nanoparticles: Nanosilica, zero-valent iron and nanoclay. *Constr. Build. Mater.* **2018**, *181*, 309–318. [CrossRef]
26. Lo, C.-K. *US 2010/0187474 A1–Pure Nanoclay and Process for Preparing Nanoclay*; Patent Application Publication: Alexandria, VA, USA, 2010.
27. Ezzat, H.; El-badawy, S.; Gabr, A.; Zaki, E.I.; Breakah, T. Evaluation of Asphalt Binders Modified with Nanoclay and Nanosilica. *Procedia Eng.* **2016**, *143*, 1260–1267. [CrossRef]
28. Tolinski, M. *Additives for Polyolefins*, 2nd ed.; William Andrew Publishing: New York, NY, USA, 2009.
29. Zimmer, A.; Jung de Andrade, M.; Sánchez, F.A.L.; Takimi, A.S. Use of Natural and Modified Natural Nanostructured Materials. In *Nanostructured Materials for Engineering Applications*; Pérez Bergmann, C., Jung de Andrade, M., Eds.; Springer: Berlin, Germany, 2011; pp. 157–172.
30. Forbes, T.Z. Occurrence of Nanomaterials in the Environment. In *Nanomaterials in the Environment*; Brar, S.K., Zhang, T.C., Verma, M., Surampalli, R.Y., Tyagi, R.D., Eds.; American Society of Civil Engineers: Reston, VR, USA, 2015; pp. 179–218.
31. Ghile, D.B. Effects of Nanoclay Modification on Rheology of Bitumen and on Performance of Asphalt Mixtures. Master's Thesis, Delft University of Technology, Mekelweg, The Netherlands, 2006.
32. Sinha Ray, S.; Okamoto, M. Polymer/layered silicate nanocomposites: A review from preparation to processing. *Prog. Polym. Sci.* **2003**, *28*, 1539–1641. [CrossRef]
33. Albdiry, M.T.; Yousif, B.F.; Ku, H.; Lau, K.T. A critical review on the manufacturing processes in relation to the properties of nanoclay/polymer composites. *J. Compos. Mater.* **2012**, *47*, 1093–1115. [CrossRef]
34. Jahromi, S.G.; Khodaii, A. Effects of nanoclay on rheological properties of bitumen binder. *Constr. Build. Mater.* **2009**, *23*, 2894–2904. [CrossRef]
35. De Melo, J.V.S.; Trichês, G. Evaluation of properties and fatigue life estimation of asphalt mixture modified by organophilic nanoclay. *Constr. Build. Mater.* **2017**, *140*, 364–373. [CrossRef]
36. Crucho, J.M.L. Development of an Accelerated Asphalt Concrete Aging Method and Utilization of Nano-Modifiers to Improve Durability of Asphalt Concrete. Ph.D. Thesis, Instituto Superior Técnico, Universidade de Lisboa, Lisboa, Portugal, 2018.
37. Crucho, J.; Neves, J.; Capitão, S.; Picado-Santos, L. Experimental study about the effects of the modification with nanoclay on the properties of a SMA mixture. In Proceedings of the TEST&E 2019–2° Congress, Porto, Portugal, 19–21 February 2019.
38. Yao, H.; You, Z.; Li, L.; Shi, X.; Goh, S.W.; Mills-beale, J.; Wingard, D. Performance of asphalt binder blended with non-modified and polymer-modified nanoclay. *Constr. Build. Mater.* **2012**, *35*, 159–170. [CrossRef]
39. Golestani, B.; Nam, B.H.; Moghadas Nejad, F.; Fallah, S. Nanoclay application to asphalt concrete: Characterization of polymer and linear nanocomposite-modified asphalt binder and mixture. *Constr. Build. Mater.* **2015**, *91*, 32–38. [CrossRef]
40. Iskender, E. Evaluation of mechanical properties of nano-clay modified asphalt mixtures. *Measurement* **2016**, *93*, 359–371. [CrossRef]
41. Zare-shahabadi, A.; Shokuhfar, A.; Ebrahimi-Nejad, S. Preparation and rheological characterization of asphalt binders reinforced with layered silicate nanoparticles. *Constr. Build. Mater.* **2010**, *24*, 1239–1244. [CrossRef]
42. Yu, J.; Zeng, X.; Wu, S.; Wang, L.; Liu, G. Preparation and properties of montmorillonite modified asphalts. *Mater. Sci. Eng. A* **2007**, *447*, 233–238. [CrossRef]
43. Goh, S.W.; Akin, M.; You, Z.; Shi, X. Effect of deicing solutions on the tensile strength of micro- or nano-modified asphalt mixture. *Constr. Build. Mater.* **2011**, *25*, 195–200. [CrossRef]
44. Wei, Y.; Fang, Z.; Zheng, L.; Tan, L.; Tsang, E.P. Green synthesis of Fe nanoparticles using Citrus maxima peels aqueous extracts. *Mater. Lett.* **2016**, *185*, 384–386. [CrossRef]

45. Wei, Y.; Fang, Z.; Zheng, L.; Tsang, E.P. Biosynthesized iron nanoparticles in aqueous extracts of Eichhornia crassipes and its mechanism in the hexavalent chromium removal. *Appl. Surf. Sci.* **2017**, *399*, 322–329. [CrossRef]
46. Oropeza, S.; Corea, M.; Gómez-Yáñez, C.; Cruz-Rivera, J.J.; Navarro-Clemente, M.E. Zero-valent iron nanoparticles preparation. *Mater. Res. Bull.* **2012**, *47*, 1478–1485. [CrossRef]
47. Wang, T.; Lin, J.; Chen, Z.; Megharaj, M.; Naidu, R. Green synthesized iron nanoparticles by green tea and eucalyptus leaves extracts used for removal of nitrate in aqueous solution. *J. Clean. Prod.* **2014**, *83*, 413–419. [CrossRef]
48. Machado, S.; Pacheco, J.G.; Nouws, H.P.A.; Albergaria, J.T.; Delerue-Matos, C. Characterization of green zero-valent iron nanoparticles produced with tree leaf extracts. *Sci. Total Environ.* **2015**, *533*, 76–81. [CrossRef] [PubMed]
49. Kordi, Z.; Shafabakhsh, G. Evaluating mechanical properties of stone mastic asphalt modified with Nano Fe2O3. *Constr. Build. Mater.* **2017**, *134*, 530–539. [CrossRef]
50. Pirmohammad, S.; Majd-Shokorlou, Y.; Amani, B. Experimental investigation of fracture properties of asphalt mixtures modified with Nano Fe2O3 and carbon nanotubes. *Road Mater. Pavement Des.* **2019**, *23*. [CrossRef]
51. Martinho, F.C.G.; Farinha, J.P.S. An overview of the use of nanoclay modified bitumen in asphalt mixtures for enhanced flexible pavement performances. *Road Mater. Pavement Des.* **2017**, *31*. [CrossRef]
52. Li, R.; Xiao, F.; Amirkhanian, S.; You, Z.; Huang, J. Developments of nano materials and technologies on asphalt materials—A review. *Constr. Build. Mater.* **2017**, *143*, 633–648. [CrossRef]
53. Fang, C.; Yu, R.; Liu, S.; Li, Y. Nanomaterials applied in asphalt modification: A review. *J. Mater. Sci. Technol.* **2013**, *29*, 589–594. [CrossRef]
54. Yang, J.; Tighe, S. A Review of Advances of Nanotechnology in Asphalt Mixtures. *Procedia Soc. Behav. Sci.* **2013**, *96*, 1269–1276. [CrossRef]
55. You, Z.; Mills-Beale, J.; Foley, J.M.; Roy, S.; Odegard, G.M.; Dai, Q.; Goh, S.W. Nanoclay-modified asphalt materials: Preparation and characterization. *Constr. Build. Mater.* **2011**, *25*, 1072–1078. [CrossRef]
56. Guo, F.; Aryana, S.; Han, Y.; Jiao, Y. A Review of the Synthesis and Applications of Polymer–Nanoclay Composites. *Appl. Sci.* **2018**, *8*, 1696. [CrossRef]
57. Polacco, G.; Filippi, S.; Merusi, F.; Stastna, G. A review of the fundamentals of polymer-modified asphalts: Asphalt/polymer interactions and principles of compatibility. *Adv. Colloid Interface Sci.* **2015**, *224*, 72–112. [CrossRef] [PubMed]
58. Yusoff, N.I.M.; Breem, A.A.S.; Alattug, H.N.M.; Hamim, A.; Ahmad, J. The effects of moisture susceptibility and ageing conditions on nano-silica/polymer-modified asphalt mixtures. *Constr. Build. Mater.* **2014**, *72*, 139–147. [CrossRef]
59. Santagata, E.; Baglieri, O.; Tsantilis, L.; Chiappinelli, G.; Aimonetto, I.B. Effect of sonication on high temperature properties of bituminous binders reinforced with nano-additives. *Constr. Build. Mater.* **2015**, *75*, 395–403. [CrossRef]
60. de Melo, J.V.S.; Trichês, G.; de Rosso, L.T. Experimental evaluation of the influence of reinforcement with Multi-Walled Carbon Nanotubes (MWCNTs) on the properties and fatigue life of hot mix asphalt. *Constr. Build. Mater.* **2018**, *162*, 369–382. [CrossRef]
61. Tsantilis, L.; Baglieri, O.; Santagata, E. Low-temperature properties of bituminous nanocomposites for road applications. *Constr. Build. Mater.* **2018**, *171*, 397–403. [CrossRef]
62. Guo, X.; Sun, M.; Dai, W.; Chen, S. Performance characteristics of silane silica modified asphalt. *Adv. Mater. Sci. Eng.* **2016**, *2016*. [CrossRef]
63. Rafi, J.; Kamal, M.; Ahmad, N.; Hafeez, M.; Faizan ul Haq, M.; Aamara Asif, S.; Shabbir, F.; Bilal Ahmed Zaidi, S. Performance Evaluation of Carbon Black Nano-Particle Reinforced Asphalt Mixture. *Appl. Sci.* **2018**, *8*, 1114. [CrossRef]
64. Nejad, F.M.; Nazari, H.; Naderi, K.; Karimiyan Khosroshahi, F.; Hatefi Oskuei, M. Thermal and rheological properties of nanoparticle modified asphalt binder at low and intermediate temperature range. *Pet. Sci. Technol.* **2017**, *35*, 641–646. [CrossRef]
65. Haq, M.; Ahmad, N.; Nasir, M.; Jamal; Hafeez, M.; Rafi, J.; Zaidi, S.; Haroon, W. Carbon Nanotubes (CNTs) in Asphalt Binder: Homogeneous Dispersion and Performance Enhancement. *Appl. Sci.* **2018**, *8*, 2651. [CrossRef]

66. Faramarzi, M.; Arabani, M.; Haghi, A.K.; Mottaghitalab, V. Carbon nanotubes-modified asphalt binder: Preparation and characterization. *Int. J. Pavement Res. Technol.* **2015**, *8*, 29–37. [CrossRef]
67. Khattak, M.J.; Khattab, A.; Rizvi, H.R.; Zhang, P. The impact of carbon nano-fiber modification on asphalt binder rheology. *Constr. Build. Mater.* **2012**, *30*, 257–264. [CrossRef]
68. European Committee for Standardization. *EN 13399. Bitumen and Bituminous Binders—Determination of Storage Stability of Modified Bitumen*; CEN: Brussels, Belgium, 2017.
69. ASTM International. *ASTM D5892-00. Standard Specification for Type IV Polymer-Modified Asphalt Cement for Use in Pavement Construction*; ASTM: West Conshohocken, PA, USA, 2000.
70. Santagata, E.; Baglieri, O.; Tsantilis, L.; Chiappinelli, G. Storage Stability of Bituminous Binders Reinforced with Nano-Additives. In *8th RILEM International Symposium on Testing and Characterization of Sustainable and Innovative Bituminous Materials*; Canestrari, F., Partl, M., Eds.; Springer: Dordrecht, The Netherlands, 2016.
71. Ramesh, K.T. *Nanomaterials*; Springer: Berlin, Germany, 2009.
72. Grassian, V.H.; O'Shaughnessy, P.T.; Adamcakova-Dodd, A.; Pettibone, J.M.; Thorne, P.S. Inhalation exposure study of Titanium dioxide nanoparticles with a primary particle size of 2 to 5 nm. *Environ. Health Perspect.* **2007**, *115*, 397–402. [CrossRef] [PubMed]
73. National Research Council. *A Research Strategy for Environmental, Health, and Safety Aspects of Engineered Nanomaterials*; The National Academies Press: Washington, DC, USA, 2012.
74. Haase, A.; Luch, A. Genotoxicity of nanomaterials in vitro: Treasure or trash? *Arch. Toxicol.* **2016**, *90*, 2827–2830. [CrossRef]
75. Ramachandran, G.; Ostraat, M.; Evans, D.E.; Methner, M.M.; O'Shaughnessy, P.; D'Arcy, J.; Geraci, C.L.; Stevenson, E.; Maynard, A.; Rickabaugh, K. A strategy for assessing workplace exposures to nanomaterials. *J. Occup. Environ. Hyg.* **2011**, *8*, 673–685. [CrossRef]
76. Wu, S.; Tahri, O. State-of-art carbon and graphene family nanomaterials for asphalt modification. *Road Mater. Pavement Des.* **2019**, *22*. [CrossRef]
77. Fini, E.; Hajikarimi, P.; Rahi, M.; Nejad, F.M. Physiochemical, Rheological, and Oxidative Aging Characteristics of Asphalt Binder in the Presence of Mesoporous Silica Nanoparticles. *J. Mater. Civ. Eng.* **2016**, *28*. [CrossRef]
78. Alhamali, D.I.; Yusoff, N.I.; Wu, J.; Liu, Q.; Albrka, S.I. The Effects of Nano Silica Particles on the Physical Properties and Storage Stability of Polymer-Modified Bitumen. *J. Civ. Eng. Res.* **2015**, *5*, 11–16. [CrossRef]
79. Onochie, A.; Fini, E.; Yang, X.; Mills-Beale, J.; You, Z. Rheological Characterization of Nano-particle based Bio-modified Binder. Transp. In Proceedings of the Transportation Research Board 92nd Annual Meeting, Washington, DC, USA, 13–17 January 2013.
80. Baldi-Sevilla, A.; Montero, M.L.; Aguiar-Moya, J.P.; Loría, L.G. Influence of nanosilica and diatomite on the physicochemical and mechanical properties of binder at unaged and oxidized conditions. *Constr. Build. Mater.* **2016**, *127*, 176–182. [CrossRef]
81. Nazari, H.; Naderi, K.; Moghadas Nejad, F. Improving aging resistance and fatigue performance of asphalt binders using inorganic nanoparticles. *Constr. Build. Mater.* **2018**, *170*, 591–602. [CrossRef]
82. Yazdani, A.; Pourjafar, S. Optimization of Asphalt Binder Modified with PP/SBS/Nanoclay Nanocomposite using Taguchi Method. *Int. Sch. Sci. Res. Innov.* **2012**, *6*, 532–536.
83. Abdullah, M.E.; Zamhari, K.A.; Hainin, M.R.; Oluwasola, E.A.; Hassan, N.A.; Yusoff, N.I.M. Engineering properties of asphalt binders containing nanoclay and chemical warm-mix asphalt additives. *Constr. Build. Mater.* **2016**, *112*, 232–240. [CrossRef]
84. Yu, J.; Wang, X.; Hu, L.; Tao, Y. Effect of Various Organomodified Montmorillonites on the Properties of Montmorillonite/Bitumen Nanocomposites. *J. Mater. Civ. Eng.* **2010**, *22*, 788–793. [CrossRef]
85. Santagata, E.; Baglieri, O.; Tsantilis, L.; Chiappinelli, G. Effects of Nano-sized Additives on the High-Temperature Properties of Bituminous Binders: A Comparative Study. In *Multi-Scale Modeling and Characterization of Infrastructure Materials, RILEM 8*; Kringos, N., Birgisson, B., Frost, D., Wang, L., Eds.; Springer: Berlin, Germany, 2013; pp. 297–309.
86. Mills-Beale, J.; You, Z. Nanoclay-Modified Asphalt Binder Systems. In *Nanotechnology in Civil Infrastructure*; Gopalakrishnan, K., Birgisson, B., Taylor, P., Attoh-Okine, N.O., Eds.; Springer: Berlin, Germany, 2011; pp. 257–270.
87. Siddig, E.A.A.; Feng, C.P.; Ming, L.Y. Effects of ethylene vinyl acetate and nanoclay additions on high-temperature performance of asphalt binders. *Constr. Build. Mater.* **2018**, *169*, 276–282. [CrossRef]

88. Golestani, B.; Moghadas Nejad, F.; Sadeghpour Galooyak, S. Performance evaluation of linear and nonlinear nanocomposite modified asphalts. *Constr. Build. Mater.* **2012**, *35*, 197–203. [CrossRef]
89. Polacco, G.; Kříž, P.; Filippi, S.; Stastna, J.; Biondi, D.; Zanzotto, L. Rheological properties of asphalt/SBS/clay blends. *Eur. Polym. J.* **2008**, *44*, 3512–3521. [CrossRef]
90. Abdelrahman, M.; Katti, D.R.; Ghavibazoo, A.; Upadhyay, H.B.; Katti, K.S. Engineering physical properties of asphalt binders through nanoclay-asphalt interactions. *J. Mater. Civ. Eng.* **2014**, *26*. [CrossRef]
91. Yao, H.; You, Z.; Li, L.; Goh, S.W.; Lee, C.H.; Yap, Y.K.; Shi, X. Rheological properties and chemical analysis of nanoclay and carbon microfiber modified asphalt with Fourier transform infrared spectroscopy. *Constr. Build. Mater.* **2013**, *38*, 327–337. [CrossRef]
92. Yao, H.; You, Z.; Li, L.; Shi, X.; Hansen, M. Performance Evaluation of Asphalt Mixtures Blended by Nonmodified Nanoclay and Polymer-Modified Nanoclay. In Proceedings of the Transportation Research Board 92nd Annual Meeting, Washington, DC, USA, 13–17 January 2013.
93. Bagshaw, S.A.; Kemmitt, T.; Waterland, M.; Brooke, S. Effect of blending conditions on nano-clay bitumen nanocomposite properties. *Road Mater. Pavement Des.* **2018**, *22*. [CrossRef]
94. Gedafa, D.S.; Landrus, D.; Suleiman, N. Effect of Nanoclay on Binder Rheology and Rutting Resistance of HMA Mixes. In Proceedings of the Transportation Research Board 96nd Annual, Washington, DC, USA, 8–12 January 2017.
95. Burguete, L.M. Contribution to the study of bituminous mixtures with nanomaterials (in Portuguese). Master's Thesis, Instituto Superior Técnico, Universidade de Lisboa, Lisboa, Portugal, 2013.
96. Sun, L.; Xin, X.; Ren, J. Asphalt modification using nano-materials and polymers composite considering high and low temperature performance. *Constr. Build. Mater.* **2017**, *133*, 358–366. [CrossRef]
97. Ezzat, H.; El-Badawy, S.; Gabr, A.; Zaki, S.; Breakah, T. Predicted performance of hot mix asphalt modified with nano-montmorillonite and nano-silicon dioxide based on Egyptian conditions. *Int. J. Pavement Eng.* **2018**, *11*. [CrossRef]
98. Tanzadeh, R.; Tanzadeh, J.; honarmand, M.; Tahami, S.A. Experimental study on the effect of basalt and glass fibers on behavior of open-graded friction course asphalt modified with nano-silica. *Constr. Build. Mater.* **2019**, *212*, 467–475. [CrossRef]
99. Van de Ven, M.; Besamusca, J.; Molenaar, A. Nanoclay for binder modification of asphalt mixtures. In Proceedings of the 7th International RILEM Symposium ATCBM09–Advanced Testing and Characterization of Bituminous Materials, Rhodes, Greece, 27–29 May 2009; Loizos, A., Partl, M.N., Scarpas, T., Al-Qadi, I.L., Eds.; pp. 133–142. [CrossRef]
100. López-Montero, T.; Crucho, J.; Picado-Santos, L.; Miró, R. Effect of nanomaterials on ageing and moisture damage using the indirect tensile strength test. *Constr. Build. Mater.* **2018**, *168*, 31–40. [CrossRef]
101. Blom, J.; de Kinder, B.; Meeusen, J.; Van den bergh, W. The influence of nanoclay on the durability properties of asphalt mixtures for top and base layers. *IOP Conf. Ser. Mater. Sci. Eng.* **2017**, *236*, 8. [CrossRef]
102. Yang, Z.; Zhang, Y.; Shi, X. Impact of nanoclay and carbon microfiber in combating the deterioration of asphalt concrete by non-chloride deicers. *Constr. Build. Mater.* **2018**, *160*, 514–525. [CrossRef]
103. Jahromi, S.G.; Andalibizade, B.; Vossough, S. Engineering properties of nanoclay modified asphalt concrete mixtures. *Arab. J. Sci. Eng.* **2010**, *35*, 89–103.
104. Ameri, M.; Mohammadi, R.; Vamegh, M.; Molayem, M. Evaluation the effects of nanoclay on permanent deformation behavior of stone mastic asphalt mixtures. *Constr. Build. Mater.* **2017**, *156*, 107–113. [CrossRef]
105. European Committee for Standardization. *EN 12591:2009. Bitumen and Bituminous Binders—Specifications for Paving Grade Bitumens*; CEN: Brussels, Belgium, 2009.
106. American Association of State Highway and Transportation Officials. *AASHTO M320-10. Standard Specification for Performance-Graded Asphalt Binder*; AASHTO: Washington, DC, USA, 2010.
107. Galooyak, S.S.; Dabir, B.; Nazarbeygi, A.E.; Moeini, A.; Berahman, B. The Effect of Nanoclay on Rheological Properties and Storage Stability of SBS-Modified Bitumen. *Pet. Sci. Technol.* **2011**, *29*, 850–859. [CrossRef]
108. Mahdi, L.M.J.; Muniandy, R.; Yunus, R.B.; Hasham, S.; Aburkaba, E. Effect of Short Term Aging on Organic Montmorillonite Nanoclay Modified Asphalt. *Indian J. Sci. Technol.* **2013**, *6*, 5434–5442.
109. Yu, J.Y.; Feng, P.C.; Zhang, H.L.; Wu, S.P. Effect of organo-montmorillonite on aging properties of asphalt. *Constr. Build. Mater.* **2009**, *23*, 2636–2640. [CrossRef]
110. Zhang, H.; Zhang, D.; Zhu, C. Properties of Bitumen Containing Various Amounts of Organic Montmorillonite. *J. Mater. Civ. Eng.* **2015**, *27*, 7. [CrossRef]

111. Ashish, P.K.; Singh, D.; Bohm, S. Investigation on influence of nanoclay addition on rheological performance of asphalt binder. *Road Mater. Pavement Des.* **2017**, *18*, 1007–1026. [CrossRef]
112. Abdullah, M.E.; Zamhari, K.A.; Buhari, R.; Nayan, M.N.; Mohd Rosli, H. Short Term and Long Term Aging Effects of Asphalt Binder Modified with Montmorillonite. *Key Eng. Mater.* **2014**, *594–595*, 996–1002. [CrossRef]
113. Ashish, P.K.; Singh, D.; Bohm, S. Evaluation of rutting, fatigue and moisture damage performance of nanoclay modified asphalt binder. *Constr. Build. Mater.* **2016**, *113*, 341–350. [CrossRef]
114. Kavussi, A.; Barghabany, P. Investigating Fatigue Behavior of Nanoclay and Nano Hydrated Lime Modified Bitumen Using LAS Test. *J. Mater. Civ. Eng.* **2015**, *28*, 1–7. [CrossRef]
115. Das, P.K.; Baaj, H.; Kringos, N.; Tighe, S. Coupling of oxidative ageing and moisture damage in asphalt mixtures. *Road Mater. Pavement Des.* **2015**, *16*, 265–279. [CrossRef]
116. Durrieu, F.; Farcas, F.; Mouillet, V. The influence of UV aging of a Styrene/Butadiene/Styrene modified bitumen: Comparison between laboratory and on site aging. *Fuel* **2007**, *86*, 1446–1451. [CrossRef]
117. Lins, V.F.C.; Araújo, M.F.A.S.; Yoshida, M.I.; Ferraz, V.P.; Andrada, D.M.; Lameiras, F.S. Photodegradation of hot-mix asphalt. *Fuel* **2008**, *87*, 3254–3261. [CrossRef]
118. Lopes, M.; Zhao, D.; Chailleux, E.; Kane, M.; Gabet, T.; Petiteau, C.; Soares, J. Characterization of aging processes on the asphalt mixture surface. *Road Mater. Pavement Des.* **2014**, *15*, 477–487. [CrossRef]
119. Mouillet, V.; Farcas, F.; Chailleux, E.; Sauger, L. Evolution of bituminous mix behaviour submitted to UV rays in laboratory compared to field exposure. *Mater. Struct.* **2014**, *47*, 1287–1299. [CrossRef]
120. Tauste, R.; Moreno-Navarro, F.; Sol-Sánchez, M.; Rubio-Gámez, M.C. Understanding the bitumen ageing phenomenon: A review. *Constr. Build. Mater.* **2018**, *192*, 593–609. [CrossRef]
121. de Sá Araújo, M.D.F.A.; Lins, V.D.F.C.; Pasa, V.M.D.; Leite, L.F.M. Weathering aging of modified asphalt binders. *Fuel Process. Technol.* **2013**, *115*, 19–25. [CrossRef]
122. Feng, Z.-G.; Bian, H.; Li, X.; Yu, J. FTIR analysis of UV aging on bitumen and its fractions. *Mater. Struct.* **2016**, *49*, 1381–1389. [CrossRef]
123. Feng, Z.-G.; Yu, J.-Y.; Zhang, H.-L.; Kuang, D.-L.; Xue, L.-H. Effect of ultraviolet aging on rheology, chemistry and morphology of ultraviolet absorber modified bitumen. *Mater. Struct.* **2013**, *46*, 1123–1132. [CrossRef]
124. Wu, S.; Pang, L.; Liu, G.; Zhu, J. Laboratory Study on Ultraviolet Radiation Aging of Bitumen. *J. Mater. Civ. Eng.* **2010**, *22*, 767–772. [CrossRef]
125. Xu, S.; Yu, J.; Zhang, C.; Sun, Y. Effect of ultraviolet aging on rheological properties of organic intercalated layered double hydroxides modified asphalt. *Constr. Build. Mater.* **2015**, *75*, 421–428. [CrossRef]
126. Yi, M.; Delehei; Wang, J.; Feng, X. Effect of Ultraviolet Light Aging on Fatigue Properties of Asphalt. *Key Eng. Mater.* **2014**, *599*, 125–129. [CrossRef]
127. Zeng, W.; Wu, S.; Wen, J.; Chen, Z. The temperature effects in aging index of asphalt during UV aging process. *Constr. Build. Mater.* **2015**, *93*, 1125–1131. [CrossRef]
128. Zhang, H.; Zhu, C.; Yu, J.; Shi, C.; Zhang, D. Influence of surface modification on physical and ultraviolet aging resistance of bitumen containing inorganic nanoparticles. *Constr. Build. Mater.* **2015**, *98*, 735–740. [CrossRef]
129. Zhang, H.; Yu, J.; Wu, S. Effect of montmorillonite organic modification on ultraviolet aging properties of SBS modified bitumen. *Constr. Build. Mater.* **2012**, *27*, 553–559. [CrossRef]
130. Zhang, H.; Yu, J.; Wang, H.; Xue, L. Investigation of microstructures and ultraviolet aging properties of organo-montmorillonite/SBS modified bitumen. *Mater. Chem. Phys.* **2011**, *129*, 769–776. [CrossRef]
131. Crucho, J.M.L.; das Neves, J.M.C.; Capitão, S.D.; de Picado-Santos, L.G. Evaluation of the durability of asphalt concrete modified with nanomaterials using the TEAGE aging method. *Constr. Build. Mater.* **2019**, *214*, 178–186. [CrossRef]
132. Crucho, J.; Picado-Santos, L.; Neves, J.; Capitão, S.; Al-Qadi, I.L. Tecnico accelerated ageing (TEAGE)—A new laboratory approach for bituminous mixture ageing simulation. *Int. J. Pavement Eng.* **2018**. [CrossRef]
133. American Association of State Highway and Transportation Officials. *AASHTO R30-02. Standard Practice for Mixture Conditioning of Hot Mix Asphaltl*; AASHTO: Washington, DC, USA, 2010.
134. Bell, C.A.; Abwahab, Y.; Cristi, M.E.; Sosnovske, D. Selection of Laboratory Aging Procedures for Asphalt-Aggregate Mixtures. In *SHRP-A-383*; Strategic Highway Research Program: Washington, DC, USA, 1994; p. 92, ISBN 0309057620.

135. Choi, Y.K. *Development of the Saturation Ageing Tensile Stiffness (SATS) Test for High Modulus Base Materials*; University of Nottingham: Nottingham, UK, 2005.
136. Steiner, D.; Hofko, B.; Hospodka, M.; Handle, F.; Grothe, H.; Füssl, J.; Eberhardsteiner, L.; Blab, R. Towards an optimised lab procedure for long-term oxidative ageing of asphalt mix specimen. *Int. J. Pavement Eng.* **2016**, *17*, 471–477. [CrossRef]
137. Martinho, F.C.G.; Picado-Santos, L.G.; Capitão, S.D. Feasibility assessment of the use of recycled aggregates for asphalt mixtures. *Sustainability* **2018**, *10*, 1737. [CrossRef]
138. Sargin, Ş.; Saltan, M.; Morova, N.; Serin, S.; Terzi, S. Evaluation of rice husk ash as filler in hot mix asphalt concrete. *Constr. Build. Mater.* **2013**, *48*, 390–397. [CrossRef]

© 2019 by the authors. Licensee MDPI, Basel, Switzerland. This article is an open access article distributed under the terms and conditions of the Creative Commons Attribution (CC BY) license (http://creativecommons.org/licenses/by/4.0/).

Article

Laboratory Assessment of Deteriorating Performance of Nano Hydrophobic Silane Silica Modified Asphalt in Spring-Thaw Season

Wei Guo [1], Xuedong Guo [1], Wuxing Chen [1], Yingsong Li [1], Mingzhi Sun [2] and Wenting Dai [1,*]

1. School of Transportation, Jilin University, Changchun 130022, China; guowei17@mails.jlu.edu.cn (W.G.); guoxd@jlu.edu.cn (X.G.); chenwx16@mails.jlu.edu.cn (W.C.); ysli16@mails.jlu.edu.cn (Y.L.)
2. Research Institute of Highway, Ministry of Transport of China, Beijing 100088, China; mzsun14@mails.jlu.edu.cn
* Correspondence: daiwt@jlu.edu.cn; Tel.: +86-130-8681-4298

Received: 13 May 2019; Accepted: 1 June 2019; Published: 4 June 2019

Abstract: In the seasonal frozen regions, freeze-thaw (F-T) damage is the main pavement damage, causing a variety of poor conditions in bitumen pavement, such as cracks, pits, potholes, and slush. In previous studies, we evaluated the effect of nano hydrophobic silane silica (NHSS) on the degradation of asphalt mixture under F-T cycles, and established the damage model of NHSS modified asphalt mixture in spring-thawing season. To gain more understanding of the influence of NHSS on asphalt in spring-thawing season, NHSS modified asphalt was systematically analyzed under F-T aging process in this study. The main research objective of this paper was to investigate the deteriorating properties of NHSS modified asphalt under Freeze-thaw aging process. Within this article, the physicochemical characteristics of NHSS modified asphalt were determined by using various laboratory tests, which included basic property test, dynamic shear rheometer test (DSR), Fourier transform infrared spectroscopy test (FTIR) and thermogravimetric analysis (TGA). The results showed that the incorporation of NHSS could inhibit the F-T aging process of asphalt. Moreover, the chemical composition and thermal stability of asphalt under F-T aging process was analyzed through FITR and TGA test parameters. The results illustrated that the sulfoxide functional groups content index was more suitable for evaluating the aging degree of asphalt in the spring-thawing season and the F-T aging process had a great impact on the thermal property of NHSS modified asphalt.

Keywords: spring-thaw season; freeze-thaw cycle; Nanomaterial modifier; nano hydrophobic silane silica; property improvement

1. Introduction

Asphalt is the most common pavement building material, and more than 94% of pavements use asphalt materials, which give a high service level for citizens [1,2]. In recent years, increased traffic-induced loading and climate change have resulted in complex conditions, which toughened the requirements for the performance of bituminous pavement materials [3,4]. Especially in the seasonal frozen regions, freeze-thaw (F-T) damage is the main type of pavement damage, causing a variety of poor conditions in asphalt pavement, such as cracks, pits, potholes, and slush. These conditions shorten the service life of asphalt pavements, increase the maintenance frequency and costs, and affect the smooth flow of traffic and transportation safety [5–7].

In cold regions, F-T damage is the most common pavement damage of asphalt pavement [8]. In spring-thaw season, the pavement surface pavement is around 0 °C, and the volume of moisture clearly fluctuates within this temperature range. During the daytime, the snow and the aggregated ice melts and enters the surface. Since the ice layer below the base layer does not melt completely, the moisture in the surface layer is difficult to discharge quickly and efficiently. When temperature falls rapidly, the residual moisture re-condenses into ice, and expansion stress is formed in the pavement structure. Moreover, the immersion and diffusion of moisture causes a strong emulsification influence on the asphalt, which leads to the degradation of asphalt adhesion property and the moisture damage of asphalt pavement. In conclusion, under the action of moisture immersion and F-T cycles, the cohesive property of the asphalt gradually decreases, and moisture at the aggregate-asphalt interface also gradually increases, which also reduces the adhesion property of asphalt. At the same time, under the action of the dynamic moisture pressure caused by repeated action of the vehicle load, the asphalt film gradually falls off the aggregate. After the micro-damage accumulates to a certain extent, the pavement damage, such as cracks, pits, potholes, and slush, gradually appear [9–14].

In recent years, many researchers have made great efforts to comprehend the influence of F-T cycles on the degradation of asphalt and asphalt mixtures [15]. Yan et al. evaluated the effects of F-T cycles on the performance of Stone Mastic Asphalt (SMA) mixtures. The test results indicated that after 20 F-T cycles, Marshall Stability (MS) values of SMA decreased between 24.4% and 56.5% [16]. Si et al. investigated the effect of F-T cycle test on the compressive strength, resilient modulus and moisture resistance on paving mixtures. The results showed that the resilient modulus and pavement structure capacity of asphalt concrete was reduced under F-T cycles [15]. Islam et al. investigated the effect of long-term F-T cycles on the stiffness and tensile strength of asphalt concrete in the laboratory. The results showed that the flexural stiffness of the F-T conditioned samples decreases with F-T conditioning, whereas the indirect tensile strength (ITS) of AC does not change significantly with F-T cycles [17]. In conclusion, F-T cycle has shown a negative effect on the engineering properties of asphalt and asphalt mixes, as the asphalt and asphalt samples have shown a decrease in strength and stiffness.

Since the generation and development of the freeze–thaw cycle damage is an extremely complex problem, the most cost-effective technique for mitigating freeze–thaw cycle damage and extending the service life of asphalt pavement is add modifiers. Nian et al. studied the influence of freeze-thaw cycles on the high-temperature characteristics of SBS-modified asphalt. The results showed that the incorporation of SBS modifiers could increase the elasticity of the bitumen significantly and slow the tendency of the bitumen's complex shear modulus to increase with the freeze-thaw cycles to a certain extent [7]. Dong et al. concluded that crumb-rubber-modified (CRM) asphalt mixture had the outstanding freeze-thaw resistance through F-T cycles [18].

Nanomaterials have a wide range of applications in many fields due to their unique production process and performance characteristics. Due to the huge surface area of inorganic nanomaterials, the incorporation of nanomaterials can increase the viscosity of the asphalt, increase the cohesion of the asphalt cement, and improve the mechanical properties of the asphalt mixtures. In recent years, many scholars have conducted research on the application of nanomaterials in the field of modified asphalt [19,20]. Hamedi et al. indicated that nano-$CaCO_3$ increased the wettability of the asphalt binder on the aggregate, promoted the aggregate-asphalt interface bonding strength and disputed the F-T damage [21]. Akbari concluded that the addition of nano-clay and nano-lime could reduce the moisture susceptibility of hot mix asphalt (HMA) and increase its durability under F-T cycles [22]. Gong et al. found that nano TiO_2, $CaCO_3$ and basalt fiber composite modified asphalt mixtures had outstanding freeze-thaw resistance by measuring the mesoscopic void volume, stability, indirect tensile stiffness modulus, splitting strength, uniaxial compression static, and dynamic creep rate [23].

In summary, the addition of nanomaterials can effectively alleviate the impact of F-T cycle on asphalt. In previous studies, we evaluated the effect of nano hydrophobic silane silica (NHSS) on the degradation of asphalt mixture under F-T cycles, and established the damage model of NHSS modified asphalt mixture in spring-thawing season. Within this framework, the main goal of this paper is to evaluate the durability and property of nano hydrophobic silane silica modified asphalt under the Freeze-thaw aging process. To the authors' knowledge, the open literature has no experimental studies of nano hydrophobic silane silica modified asphalt subjected to F-T aging process. In order to investigate the deteriorating properties of nano hydrophobic silane silica modified asphalt under

Freeze-thaw aging process systematically, the penetration, softening points, ductility test, rotational viscosity test, dynamic shear rheometer test (DSR), Fourier transform infrared spectroscopy test (FTIR) and thermogravimetric analysis test (TGA) were employed in this paper. The penetration, softening points, ductility test and rotational viscosity test are conventional physical property tests that explore the deteriorating properties of asphalt from the perspective of physical properties, and the DSR, FTIR and TGA test was applied from the perspective of rheological properties, chemical properties and thermal properties, respectively. The research plane flowchart is shown in Figure 1.

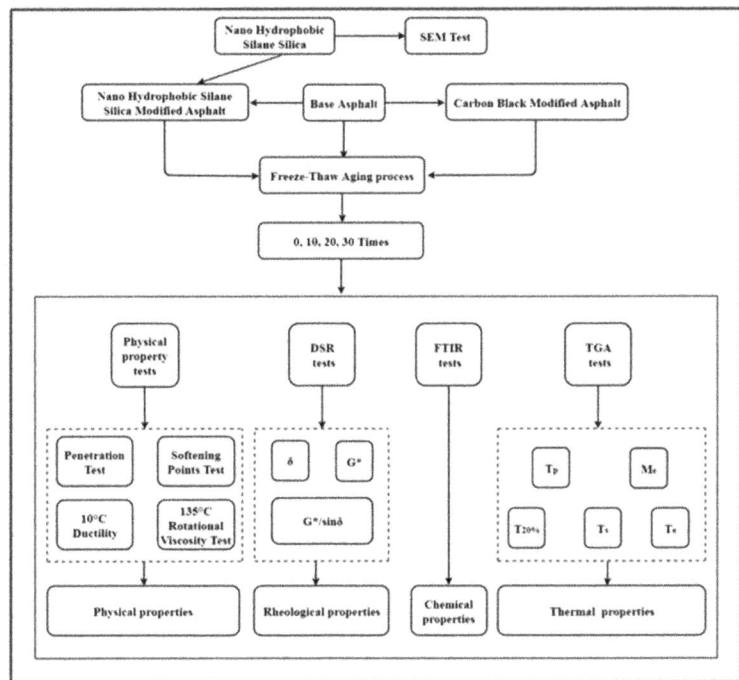

Figure 1. Research plane flowchart.

2. Materials

2.1. Asphalt

We selected 90# asphalt binder produced by Panjin Northern Asphalt Co., Ltd. for base asphalt. The asphalt binder is the most common paving grade binder used in Jilin province. The technical parameters of the asphalt used in this study are summarized in Table 1.

Table 1. Technical parameters of asphalt.

Technical Parameters	Penetration			25 °C Ductility	Softening Point	Density	Wax Content	Flash Point
	15 °C	25 °C	30 °C					
Units		0.1 mm		cm	°C	g·cm^{-3}	%	°C
Test results	25.3	86.9	142.4	>130	44.6	1.003	18	340

2.2. Nano Hydrophobic Silane Silica

Nanosilica is one of the nano-materials that have been extensively used in asphalt mixtures. However, nanosilica as a kind of inorganic non-metallic nanomaterial that is very prone to agglomeration.

In addition, the asphalt binder is an organic cementitious material formed by high molecular hydrocarbons and non-metallic derivatives of these hydrocarbons, which makes nanosilica in asphalt binder have poor dispersibility and compatibility. In order to improve the dispersion of nanosilica in organic solvents and enhance their interaction with the medium, and to broaden the application field of nanosilica, the commonly used method is to physically or chemically react the surface of the nanosilica with the surface modifier through a certain process. The nano hydrophobic silane silica is obtained by grafting the silane coupling agent onto the surface of the nanosilica to carry out surface modification. The silane coupling agent has two groups with different properties, and the chemical formula is R-Si-X. X represents hydrolysable groups such as methoxy group (CH_3O-) or ethoxy group (C_2H_5O-), which can be condensed with a hydroxyl group on the surface of nanosilica to form a siloxane bond and to be bonded to the R group. The R group can react strongly with different matrix resins or organic materials, such as vinyl, epoxy, sulfhydryl, amino, etc. [24,25].

Nano hydrophobic silane silica remains amorphous and the crystal form does not change. The coupling agent grafts the organic group to the surface of the nanoparticle by chemical action, and the bonding ability is strong, and the modification effect more obvious. The nanosilica, after surface modification, have relatively uniform particle size and when the nanosilica changes from hydrophilic to hydrophobic, the oil absorption value of nanoparticles is increased, the agglomeration phenomenon between the nanoparticles is greatly improved, and the nanoparticles are more evenly dispersed in the organic system.

The nano hydrophobic silane silica material was obtained from Changtai Weina Chemical Co., Ltd. (Shouguang, Shandong province, China). The technical properties of nano hydrophobic silane silica material have been presented in Table 2.

Table 2. Technical parameters of nano hydrophobic silane silica.

Technical Parameters	Water Characteristics	BET (m^2/g)	Loss on Drying (105 °C, 2 h, wt%)	Average Particle Size (nm)	pH Value	SiO_2 Content (%)
Test results	Hydrophilia	125 ± 20	≤0.5	12	5.0–8.0	≥99.8
Standard values	—	130 ± 30	≤3.0	≤20	3.7–6.5	≥99.8

The microstructure of nanosilica and NHSS were examined by SU8000 electronic microscopy (Tianmei.co, in Japan). The scanning electron micrographs of nanosilica and NHSS observed at magnifying power of ×1000, ×10,000 and ×100,000, which are shown in Figures 2–4. (a and b).

Figure 2. SEM images of nanosilica and nano hydrophobic silane silica (NHSS) at magnifications of ×1000. (a) nanosilica; (b) NHSS.

Appl. Sci. **2019**, *9*, 2305

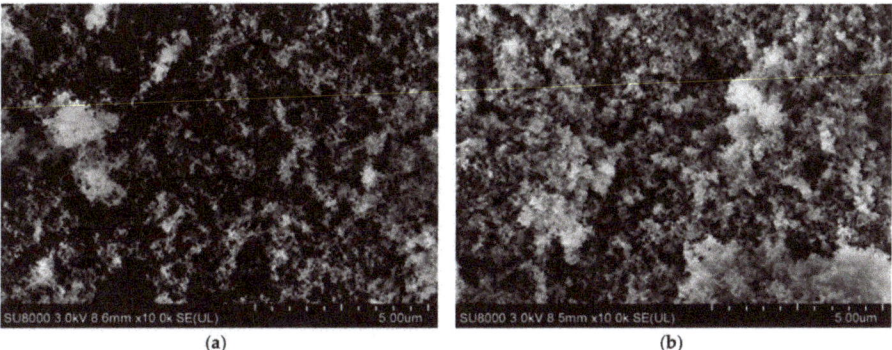

Figure 3. SEM images of nanosilica and NHSS at magnifications of ×10,000. (**a**) nanosilica; (**b**) NHSS.

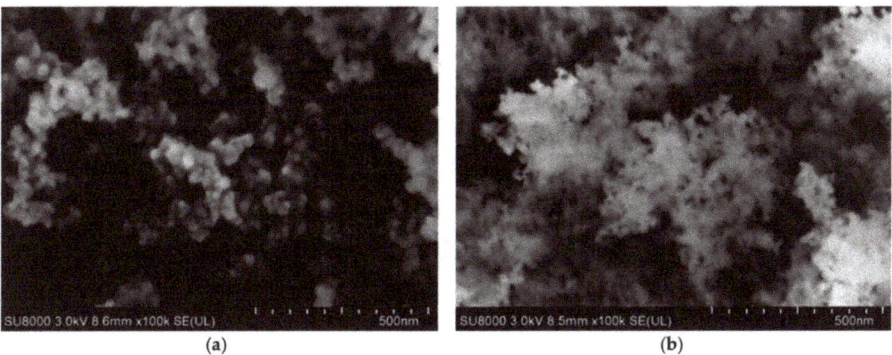

Figure 4. SEM images of nanosilica and NHSS at magnifications of ×100,000. (**a**) nanosilica; (**b**) NHSS.

The shape of nanosilica and NHSS is evident from Figure 4, and the difference between nanosilica and NHSS is more obvious. From the physical structural characteristics, the NHSS particle can better connect to asphalt. From the chemical linking characteristics, the surface modified nanosilica can form a stable chemical bond with asphalt at the interface to enhance the overall strength and toughness of the modified asphalt system.

After obtaining SEM images with different magnifications, the distribution of particle size of nanosilica and NHSS was statistically analyzed using Nano Measurer 1.2 software. The statistical results are shown in Figure 5.

The diameters of 200 particles in the nanosilica and NHSS SEM images were randomly counted. Among them, the minimum particle size of nanosilica is 28 nm, the maximum particle size is 93nm, the average particle size is 52 nm, and 91% of the particle size distribution is concentrated between 35 nm and 70 nm. NHSS has a minimum particle diameter of 15 nm, a maximum particle diameter of 82 nm, and an average particle diameter of 46 nm, 90% of the particle size distribution is concentrated in the 28 nm~69 nm range.

The dispersion of nano-scale modifiers in asphalt is an important factor limiting the development of nano-scale modified asphalt. According to Figures 4 and 5, it can be clearly concluded that the nanosilica particles are smooth, have a larger size, and are more likely to be agglomerated. This agglomeration phenomenon is mainly attributed to Van der Waals gravity and Ostwald ripening. After the surface modification of nanosilica, the particle size is slightly reduced, the surface roughness is increased, and the physical attraction between the particles is reduced, which is beneficial to improve the dispersion and fusion of NHSS particles in the asphalt, and to improve the nanosilica being easy to agglomerate.

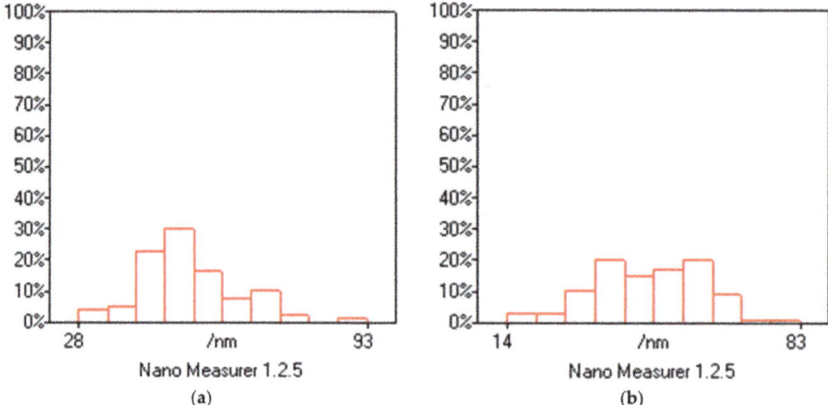

Figure 5. Nanosilica and nano hydrophobic silane silica (NHSS) particle size distribution statistics. (**a**) Nanosilica; (**b**) NHSS.

2.3. Asphalt Sample Preparation

In the previous study, the incorporation method of nano hydrophobic silane silica was discussed in detail [25]. A high shear mixer with the speed of 2000 rpm was used for incorporating the nano hydrophobic silane silica into the base asphalt. Mixing percentages of nano hydrophobic silane silica were 3 wt% of the base asphalt and the mixing temperature was kept at 140 °C. The mixing time was about 60 min to ensure homogeneous blending. The asphalt modified by nano hydrophobic silane silica was denoted by NHSSMA. Moreover, as a comparison, carbon black was selected for its unique physiochemical properties and wide application. Carbon black possesses many unique properties that distinguish it from other conventional modified: it has a large specific surface area, irregular shapes and various functional groups. Related research has proved that carbon black had good compatibility and a reinforcement effect on asphalt binders, and decreased the resistivity of asphalt [26]. In this paper, carbon black was obtained from Jiangxi black cat carbon black Co, Ltd. (Jiangxi, China). The technical information about carbon black is listed in Table 3. Thus, 3 wt% carbon black modified asphalt and base asphalt were prepared for comparison, base asphalt and carbon black modified asphalt were denoted by BA and CBMA.

Table 3. Technical parameters of carbon black.

Technical Parameters	Unit	Value
Iodine absorption	g/kg	43 ± 5
DPB absorption	10^{-5} m^3/kg	121 ± 7
DPB absorption of the compressed sample	10^{-5} m^3/kg	80~90
PH value	-	8 ± 2.0
CTAB surface area	10^3 m^2/kg	36~48
Ash content	%	≤0.7
45-μm sieve residue	mg/kg	≤1000

3. Characterization and Performance Testing

3.1. Freeze-thaw Aging Procedure

In this paper, a freeze-thaw aging process was designed to simulate the repeated effect of temperature and moisture on asphalt pavement in spring-thawing season. The freeze-thaw aging procedure of asphalt binder sample is as follows.

First, base asphalt and modified asphalt binders were heated to a fluid state and poured into a fixed-size plate to ensure the dimensions of asphalt binder samples is approximately 6 × 250 × 250 mm. The purpose of this was to ensure that the moisture could completely penetrate the asphalt and the preparation conditions of all samples were consistent.

Then, the base asphalt and modified asphalt samples were submerged in a container containing water, and the container with specimens were placed in the precision temp-enclosure at −15 °C and frozen for 10 h.

Finally, the base asphalt and modified asphalt samples were soaked in water at 15 °C for 16 h through adjusting the temperature controller.

As per the method described above, a complete freeze-thaw aging cycle was completed. Then, after 10, 20 and 30 freeze-thaw aging cycles, damaged samples were collected for physicochemical property test to explore the effect of NHSS modifier on the characteristics of asphalt under the freeze-thaw aging process. The photos of fresh and weathered specimen are shown in Figure 6.

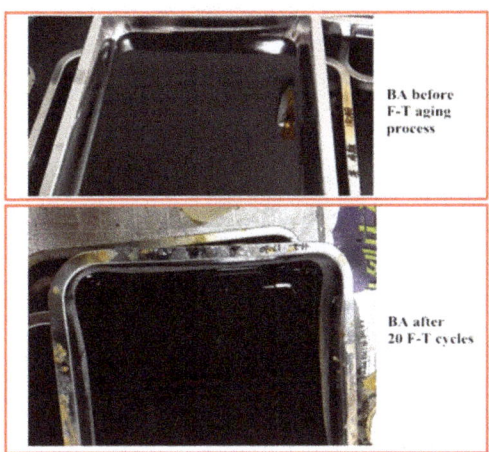

Figure 6. The photos of fresh and weathered specimen.

3.2. Property Test of the Asphalt

In order to investigate the deteriorating properties of nano hydrophobic silane silica modified asphalt under freeze-thaw aging process systematically, the penetration, softening points, ductility test, rotational viscosity test, DSR, FTIR and TGA test was employed in this paper. The penetration, softening points, ductility test and rotational viscosity test are conventional physical property tests to explore the deteriorating properties of asphalt from the perspective of physical properties, and the DSR FTIR and TGA test was applied from the perspective of rheological properties, chemical properties and thermal properties, respectively.

3.2.1. Physical Property Tests

The basic properties of the asphalt sample, including penetration, softening points and ductility, were tested according to Chinese standards GB/T4507-2010, GB/4508-2010, and GB/T4509-2010, respectively. Moreover, rotational viscosity test at 135 °C was performed according to Chinese standards GB/T0625-2011. The intercept (K) along with slope (A) were obtained to calculate Penetration Index (PI) through linearly regressing the logarithm of Penetration (P) against temperature (T).

$$PI = \frac{30}{1+50A} - 10 \tag{1}$$

The DV-III viscometer was used to measure the 135 °C rotational viscosity for evaluating the pumping ability and aging resistance of asphalt binder during F-T cycles. The aging index calculation formula based on the rotational viscosity test is as follows.

$$C = \lg\lg(\eta_a * 10^3) - \lg\lg(\eta_0 * 10^3) \quad (2)$$

where C is the aging index of the specimens, η_0 is the rotational viscosity of the specimens before freeze-thaw aging procedure, and η_a is the rotational viscosity of the specimens after different F-T cycles. The aging index reflects the upward deviation of the viscosity curve before and after freeze-thaw aging procedure. The larger the aging index value, the lesser the anti-aging ability of the asphalt. In order to ensure the repeatability of the results, three specimens were tested for each material.

3.2.2. Dynamic Shear Rheometer Test (DSR)

In order to characterize the fundamental rheological properties of asphalt film after different F-T cycles, the dynamic shear rheometer test was performed according to ASTM-D7175 standard test method. The DSR test can properly describe the elastic and viscous behaviors of asphalt film after different F-T cycles. In this paper, a Bohlin automatic dynamic shear (ADS) rheometer (DSRII, Malvern, United Kingdom) was used to investigate the rheological properties of the modified asphalt binder under freeze-thaw aging procedure. Complex shear modulus (G*) and phase angle (δ) were measured at temperatures ranging from 58 °C to 76 °C at 6 °C increments for both asphalt binders, while the frequency equaled 1.59 Hz. The parameter G* provides information about the resistance of asphalt sample to deformation when it is subjected to shear loading. The parameter (δ) shows time lag between the applied shear stresses and shear strain responses. The parameter G*/Sin δ which is called the rut factor represents the rutting resistance of asphalt sample under a freeze-thaw aging procedure.

3.2.3. Fourier Transform Infrared Spectroscopy Test (FTIR)

Fourier Transform Infrared Spectroscopy test was used to analyze the functional groups of BA, CBMA and NHSSMA under F-T aging procedure from chemical characteristics. A Vertex 70 Fourier Transform Infrared Spectroscope (Bruker Optics .co, Changchun, China) was employed with wavelength ranging from 40 cm^{-1} to 4000 cm^{-1} [27]. From the peak position and size, the chemical bonds and the functional groups of the materials in the asphalt can be determined. Based on the previous research, waves of representative chemical bonds are obtained. The results are shown in Table 4.

Table 4. Featured chemical bonds of asphalt binder.

Wave Number (cm^{-1})	Chemical Bonds
3676	Intermolecular hydrogen bond (O–H) vibration
2924	The antisymmetric stretching vibration absorption band of the alkyl (C–H)
2852	The symmetric stretching vibration absorption band of the alkyl (C–H)
1607	Conjugated double bonds (C=C) stretching vibration in aromatics
1456	The C–H asymmetric deformations in CH$_2$ and CH$_3$ vibrations
1377	The C–H symmetric deformation in CH$_3$ vibrations
1250	The C–O stretching vibration in saturated alcohols
1031	The sulfoxide group (S=O) stretching vibration
966	The C–H out-plane bending vibrations in unsaturated hydrocarbons
747	Bending vibration of aromatic branches

3.2.4. Thermogravimetric Analysis (TGA)

TGA simultaneous thermal analyzer (Netzsch .co, Bolin, Germany) was employed to measure the thermal behavior and stability properties of BA, CBMA and NHSSMA under a freeze-thaw aging procedure. The temperature range of the test was from room temperature to 900 °C, and the heating rate was controlled at 20 °C/min.

4. Results and Discussion

4.1. Basic Property Test Results

The results of basic property test are shown in Table 5. It is readily seen from this table that the 25 °C penetration of asphalt was greatly affected following the F-T aging process. From the penetration test, it can be seen that the 25 °C penetration of unmodified and modified asphalt under F-T aging progress is generally reduced and the decay rate of penetration decreases with the increase of F-T cycles, which may be due to the aging of the asphalt under the F-T aging process. With the development of F-T aging process, the proportion of asphaltenes in asphalt gradually increases, the outer membrane of the micelles becomes thinner, the mutual attraction between the micelles increases, so that the asphalt materials gradually harden, and the penetration gradually decreases.

Table 5. The results of basic property test.

Materials	F-T Cycl-es	Penetration			PI Value-	Softening Point °C	10 °C Ductility mm	Rotational Viscosity mpa	Aging Index -
		15 °C	25 °C	30 °C					
			0.1 mm						
CA	0	25.3	86.9	142.4	−1.493	44.6	1090	497.7	0
	10	19.8	57	110	−1.319	45.8	584	500	1.5×10^{-4}
	20	19	56.6	98	−1.109	46.6	504	522.3	1.59×10^{-3}
	30	18.6	54.2	96.3	−1.103	46.6	322	577.6	4.9×10^{-3}
NHSSMA	0	20.8	64.1	115	−1.357	48.1	233	912.1	0
	10	16.2	49.5	94	−1.497	49.7	172	938.6	9×10^{-4}
	20	15.1	48.9	81	−1.301	50.4	165	940.2	9.6×10^{-4}
	30	14	46.4	74	−1.273	50.5	124	945.1	1.12×10^{-3}
CBMA	0	23.9	74.8	120.5	−1.075	46.5	792	710	0
	10	18.8	59	96	−1.119	47.4	518	762.3	2.28×10^{-3}
	20	17.2	57.5	83.7	−1.012	46.5	420	771	2.65×10^{-3}
	30	16.2	54.7	78.6	−1.010	47.3	298	784.8	3.21×10^{-3}

According to the trend of penetration index in Figure 7, it can be found that NHSSMA and CBMA is less sensitive to the F-T aging process than base asphalt, and shows better temperature stability. The difference of PI development trend between modified asphalt and unmodified asphalt is the PI of CBMA and NHSSMA first decreases and then increases with the development of F-T aging process. Moreover, the PI variation of CBMA and NHSSMA is less than that of BA after 30 F-T cycles, which is due to the incorporation of carbon black modified and NHSS. After 10 F-T cycles, the previously formed joint structure between asphalt and NHSS particles was destroyed, resulting in a decrease in the PI value of asphalt. After 30 F-T cycles, the PI value increased due to the aging of the asphalt in the modified asphalt. The penetration index variation of NHSSMA was affected by the interaction of above two conditions under the F-T aging process, which lead to the difference of PI development trends between NHSSMA and BA.

The results of softening point test for BA, CBMA and NHSSMA under the F-T aging process are given in Figure 8. It is can be seen from this figure that the softening point of asphalt were greatly affected following the incorporation of modifiers. From the softening point test, it was observed that the softening points of BA, CBMA and NHSSMA increases with the development of F-T cycles. This indicates that the F-T aging progress is beneficial to the high-temperature performance of asphalt. It can also be concluded that the softening point variation of modified asphalt is much less than unmodified asphalt after 30 F-T cycles, and modified asphalt is not sensitive to F-T aging progress, which is consistent with the penetration test results.

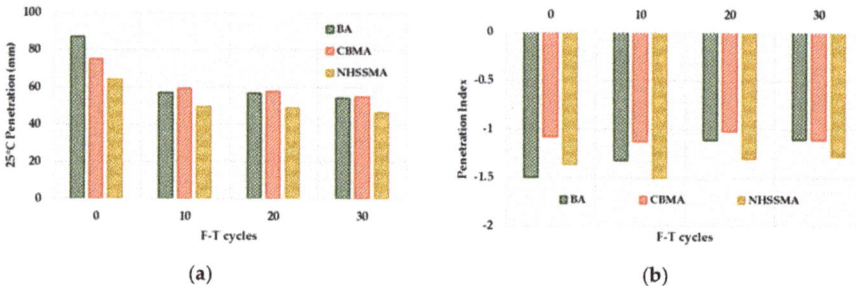

Figure 7. Penetration test results of modified and unmodified asphalt at different F-T cycles. (**a**) 25 °C penetration; (**b**) penetration index.

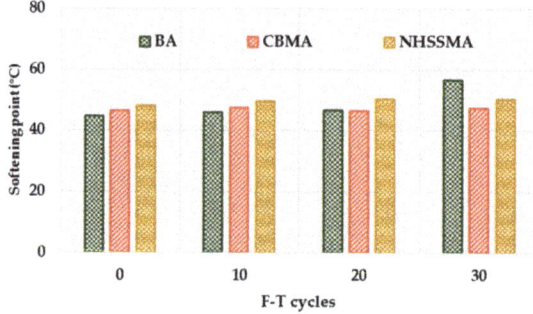

Figure 8. Softening point test results of modified and unmodified asphalt at different freeze-thaw (F-T) cycles.

The results of ductility test for BA, CBMA and NHSSMA under the F-T aging process are given in Figure 9. It can be seen from Figure 9 that the ductility of all asphalt gradually decreases with the increase of F-T cycles. This is due to the aging of the asphalt caused by F-T aging progress. As the aging process intensifies, the asphalt materials gradually harden, and the ductility of asphalt gradually decreases. The ductility of BA is greater than that of CBMA and NHSSMA under the F-T aging progress, which may be due to the destruction of homogeneity of the asphalt after the incorporation of particulate powder modifier. NHSSMA has the smallest ductility among the three types of asphalt, indicating that NHSSMA is more brittle.

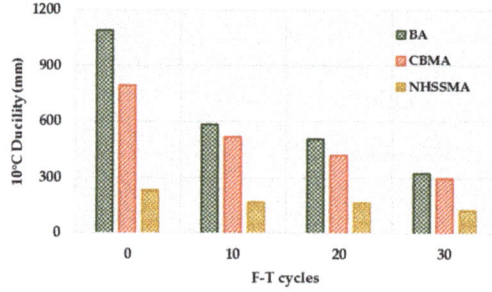

Figure 9. Ductility test results of modified and unmodified asphalt at different F-T cycles.

As can be observed, 135 °C rotational viscosity and aging index for modified and unmodified asphalt under the F-T aging process are given in Figure 10. It can be seen that 135 °C rotational viscosity and aging index for modified and unmodified asphalt increases with the increase of F-T cycles. After 30 F-T cycles, the aging index of BA, CBMA and NHSSMA is 0.0049, 0.00321 and 0.0012, respectively. NHSSMA has the lowest aging index after 30 F-T cycles, indicating that the addition of NHSS can effectively resist F-T aging progress.

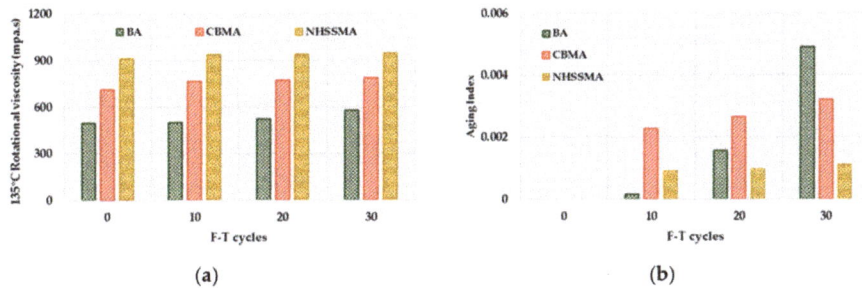

Figure 10. Rotational viscosity test results of modified and unmodified asphalt at different F-T cycles. (**a**) 135 °C rotational viscosity; (**b**) aging index.

Based on the basic properties tests such as penetration, softening point, 10 °C ductility and 135 °C rotational viscosity, it can be concluded that the incorporation of NHSS results in a relatively more stable binder compared to BA and CBMA, which may be beneficial for F-T aging resistance.

4.2. DSR Test Results

The temperature sweep tests of BA, CBMA and NHSSMA under the F-T aging process were carried out, and the complex shear modulus (G^*), phase angle (δ), rutting factor and $G^*/\sin\delta$ were analyzed.

It can be seen from Figure 11, with the increase of temperature, the complex shear modulus of BA, CBMA and NHSSMA gradually decreases, but the phase angle shows an upward trend with the increase of temperature. This is because the increase in temperature causes the volume of free asphalt to increase, and the elastic state of asphalt at low temperature gradually shifts to the flow state at high temperature. The maximum shear stress of asphalt samples decreases, and the maximum shear strain increases gradually with increasing temperature in the dynamic shear test, so the complex shear modulus of BA, CBMA and NHSSMA gradually decreases with increasing temperature. Meanwhile, as the temperature increases, the viscous component in the asphalt increases and the elastic component decreases, so the phase angle increases with increasing temperature.

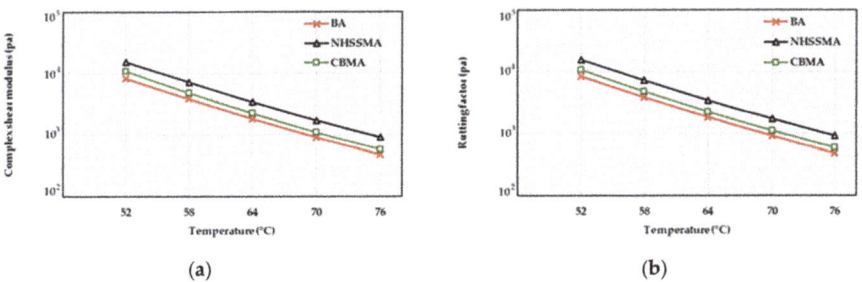

Figure 11. Complex shear modulus and rutting factor of BA, CBMA and NHSSMA without F-T aging process. (**a**) complex shear modulus; (**b**) rutting factor.

The complex modulus and the rutting factor variation tendency of BA, CBMA and NHSSMA with temperature is basically the same, which indicates that the addition of carbon black and NHSS does not change the viscoelastic properties of asphalt. However, the addition of carbon black and NHSS changes the value of complex shear modulus and rutting factor. The complex shear modulus of CBMA and NHSSMA is increased by 22~27% and 82~92% compared with BA respectively, while the increase in rutting factor of CBMA and NHSSMA is 21~28% and 45~48%. The results show that the addition of NHSS provided a certain rheological resistance of asphalt, and NHSS could better increase the viscosity of asphalt binder relative to carbon black due to the NHSS particles having higher affinity with asphalt binder functional groups through surface attraction, which is consistent with the conclusions of the basic properties test. The rutting factor of NHSSMA has a certain improvement compared with BA and CBMA, which manifested that the incorporation of NHSS improves the high temperature stability and resistance to high temperature permanent deformation of asphalt.

Figures 12 and 13 show the relationship between DSR parameters of BA and NHSSMA and F-T cycles. The changes of the rutting factor of BA and NHSSMA were relatively consistent with the changes of complex shear modulus. The rutting factor and complex shear modulus have increased under the F-T aging process. After 10 F-T cycles, the rutting factor of BA increased by 13~21% at different temperature, and the increase range was 31~34% after 20 F-T cycles. After 30 cycles, the improvement rate reached 43~54%. This is because the F-T aging progress makes the asphalt harder, reduces the fluidity, and improves the rutting factor of asphalt. After 10, 20 and 30 F-T cycles, the rutting factor of NHSSMA increased by 8~30%, 13~30% and 18~53%, indicating that the addition of NHSS can inhibit the aging effect of the F-T cycle on asphalt.

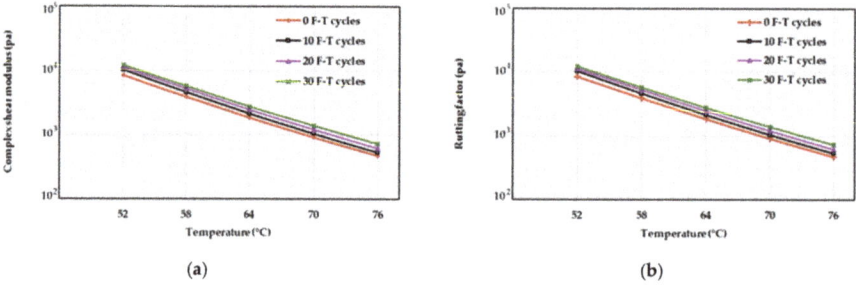

Figure 12. Complex shear modulus and rutting factor of BA at different F-T cycle. (**a**) complex shear modulus; (**b**) rutting factor.

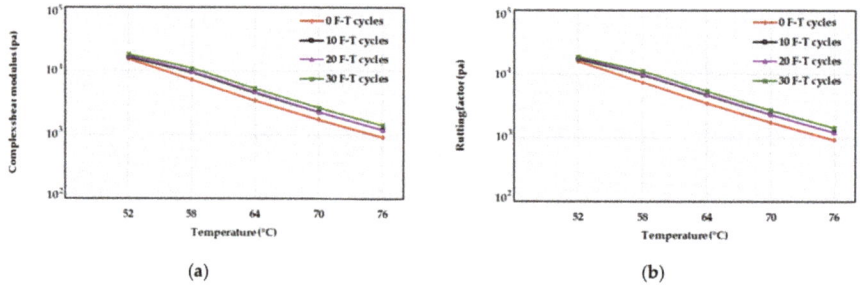

Figure 13. Complex shear modulus and rutting factor of NHSSMA at different F-T cycle. (**a**) complex shear modulus; (**b**) rutting factor.

4.3. FTIR Test Results

The FTIR spectra of the BA, CBMA and NHSSMA are given in Figures 14–16. From the FTIR spectrum of BA, it can be seen that the strong absorption peak at 2918 cm^{-1} and 2851 cm^{-1} is the asymmetric and symmetric stretching vibration absorption band of C–H in aliphatic (mainly alkanes and naphthenes), and the peak at 1600 cm^{-1} is the C–C stretching vibration in aromatics, peaks at 1458 cm^{-1} and 1375 cm^{-1} is the in-plane bending vibration absorption peak of C–H in aliphatic, peak at 1032 cm^{-1} is the sulfoxide group (S=O) stretching vibration, four small absorption peaks in the range of 863 cm^{-1}~724 cm^{-1} are the out-of-plane bending vibration of C–H on aromatic benzene rings. Thus, asphalt is mainly composed of saturated hydrocarbons, aromatics, aliphatics and heteroatom derivatives.

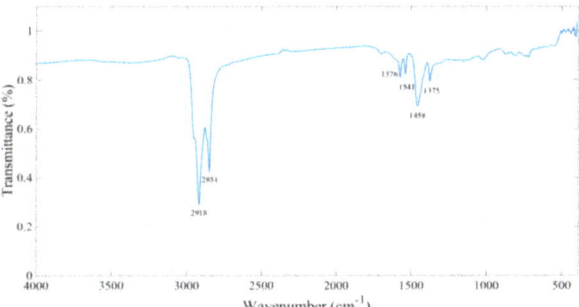

Figure 14. Fourier transform infrared spectroscopy test (FTIR) spectra of BA without F-T aging process.

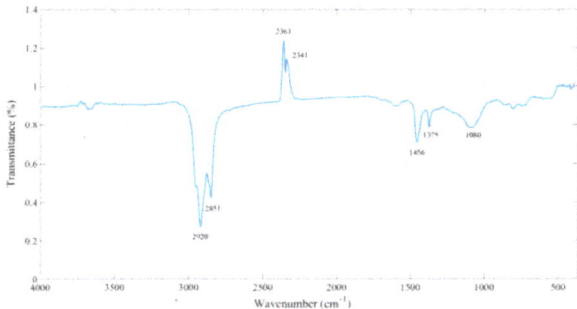

Figure 15. FTIR spectra of CBMA without F-T aging process.

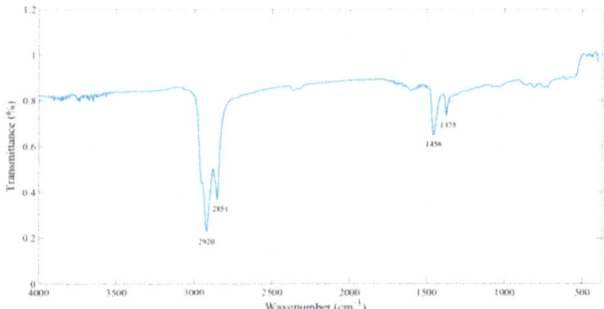

Figure 16. FTIR spectra of NHSSMA without F-T aging process.

From Figure 15, the positions and intensity of each peak in the FTIR spectra of CBMA is basically the same as that of BA, which indicates that the incorporation of carbon black does not change the properties of the asphalt, but pure physical blending.

It can be seen from Figure 16, two peaks emerged at 3675 cm^{-1} and 1300 cm^{-1}, which indicates that there is a certain chemical reaction between NHSS modifier and the asphalt except for physical blending. The strong absorption peak at 3675 cm^{-1} is the vibration absorption band of O–H in carboxyl group, which is formed by the graft reaction of nano hydrophobic silane silica with asphalt. The peak at 1300 cm^{-1} is caused by the infiltration of CO_2 during the mixing process.

After different F-T cycles, damaged specimens were collected for Fourier Transform Infrared Spectroscopy test to identify the utility of NHSS. From the FTIR spectrum of the three kinds of asphalt after different F-T cycles, it can be found that the F-T aging process does not change the peak position but changes the peak area and peak height in the FTIR spectrum.

Generally, the change of the aliphatic compounds is characterized by the aliphatic functional groups (1377 cm^{-1} and 1462 cm^{-1}). The larger the aromatic functional group content, and the higher the saturated fraction content, the greater the penetration and the lower the viscosity of asphalt. The aromatic functional group (1600 cm^{-1}) corresponds to the aromatic ring component (aromatic, colloidal and asphaltene) in the asphalt. The higher the aromatic functional group content, the higher the ductility of asphalt. Moreover, the content of carboxyl and sulfoxide functional groups is generally used as evaluation indexes to evaluate the aging degree of asphalt. The main peak area and peak height of BA, CBMA and NHSSMA under the F-T aging procedure are summarized by Origin 9.0 software in Tables 6 and 7 and Figure 17.

Table 6. The main peak area of BA, CBMA and NHSSMA under the F-T aging procedure.

Materials	F-T Cycles	3670 cm^{-1}	2924 cm^{-1}	2852 cm^{-1}	1600 cm^{-1}	1456 cm^{-1}	1377 cm^{-1}	1250 cm^{-1}	1032 cm^{-1}
CA	0 F-T cycle	0.03	35.89	18.44	2.037	9.237	1.468	0.02	0.915
	10 F-T cycles	0.09	43.97	21.33	2.311	8.145	2.528	0.239	0.96
	20 F-T cycles	4.53	62.3	23.33	1.952	7.3	1.713	1.498	8.29
	30 F-T cycles	7.43	64.55	18.35	0.792	4.496	0.829	2.153	12.21
NHSSMA	0 F-T cycles	0.83	43.42	20.87	1.387	7.44	2.409	0.067	13.72
	10 F-T cycles	0.84	51.32	22.1	1.949	6.79	1.494	0.234	12.62
	20 F-T cycles	4.93	55	23.5	1.762	7.07	1.765	0.78	14.37
	30 F-T cycles	6.11	64.86	22.75	2.436	8.03	2.79	0.873	16.91
CBMA	0 F-T cycles	0.243	42.4	21.44	2.129	7.19	1.899	0.117	1.12
	10 F-T cycles	3.33	52.3	22.07	1.65	6.69	1.856	0.948	4.87
	20 F-T cycles	2.64	55.6	22.85	1.255	6.55	1.447	1.343	6.39
	30 F-T cycles	2.63	52.5	21.89	1.21	6.36	1.821	1.116	5.05

Table 7. The main peak height of BA, CBMA and NHSSMA under the F-T aging procedure.

Materials	F-T Cycles	3670 cm^{-1}	2924 cm^{-1}	2852 cm^{-1}	1600 cm^{-1}	1456 cm^{-1}	1377 cm^{-1}	1250 cm^{-1}	1032 cm^{-1}
CA	0 F-T cycles	0.001	0.588	0.45	0.082	0.235	0.076	0.001	0.022
	10 F-T cycles	0.004	0.693	0.55	0.038	0.203	0.093	0.016	0.023
	20 F-T cycles	0.086	0.71	0.54	0.036	0.2	0.098	0.04	0.115
	30 F-T cycles	0.125	0.64	0.43	0.018	0.146	0.064	0.054	0.178
NHSSMA	0 F-T cycles	0.023	0.61	0.45	0.026	0.195	0.095	0.006	0.119
	10 F-T cycles	0.022	0.64	0.48	0.033	0.188	0.086	0.014	0.114
	20 F-T cycles	0.069	0.7	0.54	0.036	0.2	0.096	0.027	0.128
	30 F-T cycles	0.075	0.75	0.49	0.041	0.21	0.102	0.029	0.131
CBMA	0 F-T cycles	0.033	0.61	0.46	0.03	0.2	0.095	0.005	0.017
	10 F-T cycles	0.065	0.64	0.48	0.028	0.19	0.094	0.026	0.064
	20 F-T cycles	0.072	0.65	0.5	0.026	0.19	0.087	0.033	0.087
	30 F-T cycles	0.076	0.62	0.46	0.024	0.19	0.094	0.027	0.072

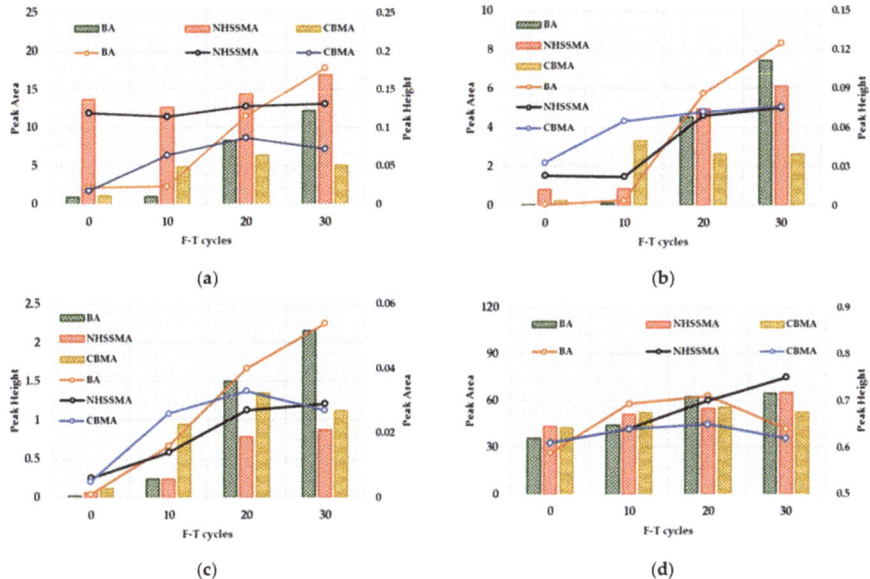

Figure 17. The changes of main functional groups of BA, CBMA and NHSSMA under the F-T aging procedure. (**a**) Sulfoxide group S=O (1032 cm^{-1}); (**b**) Free hydrocarbon group O–H (3676 cm^{-1}); (**c**) Aromatic acid group (1250 cm^{-1}); (**d**) Methylene C–H (2852 cm^{-1}).

It can be clearly seen from Figure 17 that the peak area and peak height of the main functional groups for BA, CBMA and NHSSMA generally increase as the number of F-T cycles increase. Furthermore, the sulfoxide functional groups content index of BA, CBMA, and NHSSMA changes significantly under the F-T aging process. This is mainly because the polar molecules easily react with the organic sulfides of asphalt to form sulfoxide functional groups (S=O, 1032 cm^{-1}) under the F-T aging process. Since the chemical activity of sulfur is higher than that of carbon, sulfur can participate in the reaction more rapidly under the F-T aging process. Moreover, the formation reaction of carbonyl (C=O, 1700 cm^{-1}) and the formation reaction of sulfoxide group have a competitive relationship, so when the change in the amount of the sulfoxide group is large, the amount of change in the carbonyl group is relatively small.

In addition, the sulfur content of Panjing AH-90# asphalt is relatively low compared with other asphalt, but a large number of sulfoxide groups were formed, and the carbonyl content changed very little under the F-T aging process. This may be attributed to the F-T aging process. The asphalt was kept in a freeze-thaw environment for a long time, which reduced the contact of the asphalt with the oxygen in the air and hindered the formation of carbonyl groups. In summary, the sulfoxide functional groups content index is more suitable for evaluating the aging degree of asphalt in the spring-thawing season.

The free hydroxyl groups (O–H, 3676 cm^{-1} and C–O, 1250 cm^{-1}) and methylene groups (C–H, 2852 cm^{-1}) also have a significant increase under the F-T aging process. The increase of free hydroxyl groups is closely related to the long-term diffusion of moisture in the asphalt. The intrusion of moisture accelerates the emulsification reaction of the asphalt, so the free hydrocarbon group plays an important role in evaluating the moisture damage process of the asphalt.

The change of the peak area and peak height of the main functional groups of CBMA is similar to that of BA under the F-T aging process, but the variation range is smaller than that of BA, which indicates that the incorporation of carbon black can inhibit the F-T aging process of asphalt. The NHSSMA exhibits different characteristics from the other two asphalt materials. The peak area and peak height of main functional group did not change significantly at different F-T cycles, and the distribution was stable. This indicates that NHSSMA is insensitive to the F-T aging process and has highest property stability.

4.4. TGA Test Results

TGA test was employed to evaluate the effect of NHSS on the thermal properties of asphalt under the F-T aging process. Figure 18 presents the thermal properties of modified and unmodified asphalt binder without F-T aging process. As observed, TGA thermographs relate to particular phases of degradation with distinct initial and final degradation temperatures and the addition of modifiers changes the distinct initial and final degradation temperatures, which are the result of pyrolysis of the specimen. The TGA curves of CBMA and NHSSMA are more gradual than that of BA, indicating that the addition of modifiers effectively improves the thermal properties of asphalt.

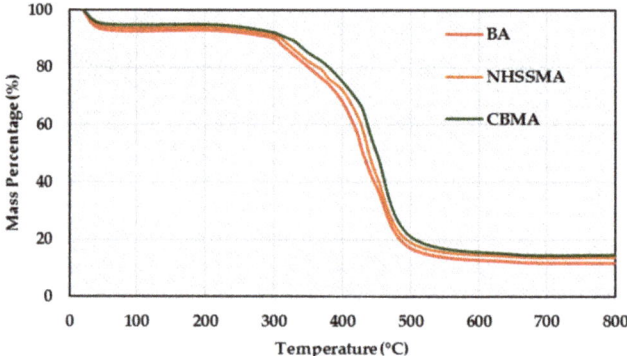

Figure 18. The thermogravimetric analysis (TGA) test results of BA, CBMA and NHSSMA without F-T aging process.

From TGA curves of all the asphalt samples, the following parameters were calculated for further comprehension of the effect of NHSS on the thermal properties of asphalt under the F-T aging process: (1) Start of thermal degradation (T_s, °C); (2) Temperature at 20% weight loss ($T_{20\%}$, °C); (3) Peak temperature (temperature at 50% weight loss) (T_p, °C); (4) End of thermal degradation (T_e, °C); (5) Residual mass (M_e, %). Table 8 and Figure 19 present a summary of TGA test parameters of BA, CBMA and NHSSMA at different F-T cycles.

Table 8. The TGA test parameters of BA, CBMA and NHSSMA at different F-T cycles.

Sample	F-T Cycles	T_S (°C)	$T_{20\%}$ (°C)	T_p (°C)	T_e (°C)	M_e (%)
CA	0	296.8	350.5	431	494.5	11.87
	10	296.1	349	428.4	493.2	9.83
	20	286.2	342.6	427.2	494.6	9.63
	30	291.1	345.3	426.6	493.8	7.45
NHSSMA	0	322.8	369.3	439	495	13.86
	10	280.1	339.4	428.3	495.4	12.08
	20	283.0	342	430.5	494.2	11.59
	30	257.0	322.3	420.2	494.6	7.60
CBMA	0	345.8	385.3	444.6	496.6	14.62
	30	293.2	358.1	432.5	496.8	10.12

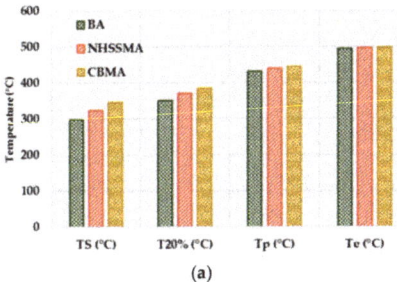 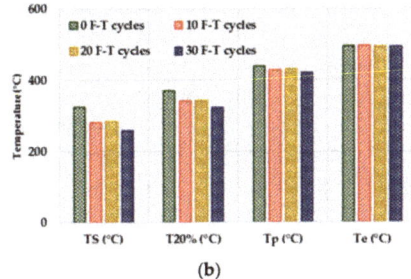

Figure 19. TGA test parameters of BA, CBMA and NHSSMA with different F-T cycles. (**a**) TGA test parameters of asphalt binders without F-T cycles; (**b**) TGA test parameters of NHSSMA at different F-T cycles.

As shown in Table 8, the T_s value of CBMA and NHSSMA increased by 49 °C and 46 °C, respectively, compared to BA, and the T_p value of CBMA and NHSSMA were 13 °C and 8 °C higher than BA. This indicates that the addition of modifiers effectively improved the thermal properties of asphalt. The M_e value of BA, CBMA and NHSSMA was 11.8%, 14.62% and 13.86%, respectively. The M_e value of modified asphalt materials was generally 2% to 3% higher than that of unmodified asphalt, which was roughly equal to the amount of the modifiers. This also indirectly indicates the modified asphalt prepared by the high-speed shearing method has good uniformity, and the modifier materials can be well dispersed in the asphalt, which can improve the storage stability of the modified asphalt.

It can be seen from Figure 19b that the T_s, $T_{20\%}$, T_p, and M_e index of NHSSMA decreased with the increase of F-T cycles. After 30 F-T cycles, the T_s value of BA, CBMA and NHSSMA decreased by 1.9%, 15.2% and 20.4%, the $T_{20\%}$ value of BA, CBMA and NHSSMA decreased by 1.5%, 7% and 12.7%, and the T_p value of BA, CBMA and NHSSMA decreased by 1%, 4% and 2.7%. This indicates that the TGA parameter variation of the modified asphalt was larger than that of BA. This may be because carbon black and NHSS are inorganic materials, and its connection with asphalt is more likely to be destroyed under the F-T aging process. The decomposition temperature of saturated fraction, aromatic fraction, colloid and asphaltene of asphalt was 300 °C, 412 °C, 438 °C, 472 °C, respectively. When the temperature reached the thermogravimetric termination temperature (T_e), the other three components, except the asphaltenes, were basically burnt out. Moreover, the base asphalt used in the test was the same asphalt, and the proportion of the four components of each asphalt sample was basically the same. Therefore, the F-T aging process and the modified material do not affect the T_e parameter.

5. Conclusions

In this paper, the property of nano hydrophobic silane silica modified asphalt in spring-thawing season was assessed according to the basic property test, DSR test, FTIR test and TGA test. The main conclusions are as follows:

1. The basic properties test results showed that F-T aging process results in the significant aging process of asphalt; it makes the asphalt more brittle and harder, increases the viscosity and softening point of the asphalt, and reduces the penetration and ductility of the asphalt. Moreover, the addition of NHSS can effectively reduce the sensitivity of asphalt to F-T cycle.
2. The DSR test results showed that the rutting factor and complex modulus of modified and unmodified asphalt gradually increase with the increase of F-T cycle. In addition, the complex modulus and rutting factor variation of NHSSMA is smaller than that of BA, indicating that the incorporation of NHSS can inhibit the F-T aging process of asphalt.
3. From the FITR test, it was understood that the sulfoxide functional groups content index is more suitable for evaluating the aging degree of asphalt in the spring-thawing season. Moreover, the peak area and peak height of main functional group for NHSSMA did not change significantly

at different F-T cycles compared to BA and CBMA, which indicated that NHSSMA was insensitive to the F-T aging process and had the highest property stability.
4. The TGA test of BA, CBMA and NHSSMA was carried out, and the influence of F-T cycle on the thermal stability of the asphalt was quantitatively based on the TGA parameters. It is found that NHSS can improve the thermal stability of asphalt, but the F-T aging process has a great influence on the thermal property of NHSSMA. This may due to the fact that NHSS is an inorganic material, and its connection with asphalt is more likely to be destroyed under F-T aging process

Author Contributions: Data curation, W.G., W.C., Y.L., M.S. and W.D.; Funding acquisition, X.G. and W.D.; Investigation, W.G. and W.C.; Methodology, W.G.; Project administration, X.G.; Writing—original draft, W.G. and Y.L.; Writing—review & editing, X.G. and W.D.

Funding: This research was funded by the National Nature Science Foundation of China (NSFC) (Grant No. 51178204).

Conflicts of Interest: The authors declare that there are no conflict of interests regarding the publication of this paper.

References

1. Shun, B.A.; Zhang, L.; Wu, S.P.; Dong, L.J.; Liu, Q.T.; Wang, Q. Synthesis and characterization of compartmented ca-alginate/silica self-healing fibers containing bituminous rejuvenator. *Constr. Build. Mater.* **2018**, *190*, 623–631. [CrossRef]
2. Agzenai, Y.; Pozuelo, J.; Sanz, J.; Perez, I.; Baselga, J. Advanced self-healing asphalt composites in the pavement performance field: Mechanisms at the nano level and new repairing methodologies. *Recent Pat. Nanotechnol.* **2015**, *9*, 43–50. [CrossRef] [PubMed]
3. Yang, Q.L.; Liu, Q.; Zhong, J.; Hong, B.; Wang, D.W.; Oeser, M. Rheological and micro-structural characterization of bitumen modified with carbon nanomaterials. *Constr. Build. Mater.* **2019**, *201*, 580–589. [CrossRef]
4. Liu, J.; Yan, K.Z.; Liu, J.; Guo, D. Evaluation of the characteristics of Trinidad Lake Asphalt and Styrene-Butadiene-Rubber compound modified binder. *Constr. Build. Mater.* **2019**, *202*, 614–621. [CrossRef]
5. Cui, P.; Shao, M.; Sun, L. Seismic stratigraphy of the quaternary Yellow River delta, Bohai Sea, eastern China. *Mar. Geophys. Res.* **2008**, *8*, 27–42. [CrossRef]
6. Molenaar, A.A.A.; Hagos, E.T.; van de Ven, M.F.C. Effects of aging on the mechanical characteristics of bituminous binders in PAC. *J. Mater. Civ. Eng.* **2010**, *22*, 779–787. [CrossRef]
7. Nian, T.F.; Li, P.; Wei, X.Y.; Wang, P.H.; Li, H.S.; Guo, R. The effect of freeze-thaw cycles on durability properties of SBS-modified bitumen. *Constr. Build. Mater.* **2018**, *187*, 77–88. [CrossRef]
8. Badeli, S.; Carter, A.; Dore, G. Complex modulus and fatigue analysis of asphalt mix after daily rapid freeze-Thaw cycles. *J. Mater. Civ. Eng.* **2018**, *30*. [CrossRef]
9. Badeli, S.; Carter, A.; Dore, G. Effect of laboratory compaction on the viscoelastic characteristics of an asphalt mix before and after rapid freeze-thaw cycles. *Cold Reg. Sci. Technol.* **2018**, *146*, 98–109. [CrossRef]
10. Si, W.; Ma, B.; Xiao, N.; Gesang, Z.R. Analysis on compression characteristics of asphalt mixture under freeze-thaw cycles in cold plateau regions. *J. Highw. Trans. Res. Dev.* **2013**, *30*, 44–48. [CrossRef]
11. Feng, D.C.; Yi, J.Y.; Wang, D.S.; Chen, L.L. Impact of salt and freeze-thaw cycles on performance of asphalt mixtures in coastal frozen region of China. *Cold Reg. Sci. Technol.* **2010**, *62*, 34–41. [CrossRef]
12. Kettil, P.; Engstrom, G.; Wiberg, N.E. Coupled hydro-mechanical wave propagation in road structures. *Comput. Struct.* **2005**, *83*, 21–22. [CrossRef]
13. Zhang, Z.; Damnjanovi, L. Applying method of moments to model reliability of pavements infrastructure. *Transp. Eng. J. ASCE* **2006**, *132*, 416–424. [CrossRef]
14. Hong, F.; Prozzj, J.A. Estimation of pavement performance deterioration using Bayesian approach. *J. Infrastruct. Syst.* **2006**, *12*, 77–86. [CrossRef]
15. Si, W.; Ma, B.; Li, N.; Ren, J.P.; Wang, H.N. Reliability-based assessment of deteriorating performance to asphalt pavement under freeze-thaw cycles in cold regions. *Constr. Build. Mater.* **2014**, *68*, 572–579. [CrossRef]
16. Yang, K.Z.; Ge, D.D.; You, L.Y.; Wang, X.L. Laboratory investigation of the characteristics of SMA mixtures under freeze-thaw cycles, *Cold Reg. Sci. Technol.* **2015**, *119*, 68–74. [CrossRef]

17. Islam, M.R.; Tarefder, R.A. Effects of Large Freeze-Thaw Cycles on Stiffness and Tensile Strength of Asphalt Concrete. *J. Cold Reg. Eng.* **2016**, *30*, 06014006. [CrossRef]
18. Dong, Y.M.; Tan, Y.Q.; Yang, L.Y. Evaluation of performance on crumb-rubber-modified asphalt mixture. *J. Test. Eval.* **2012**, *40*, 1089–1093. [CrossRef]
19. Feldman, D. Polymer nanocomposites in building, construction. Journal of macromolecular science part a-pure and applied chemistry. *J. Macromol. Sci. Part A* **2014**, *51*, 203–209. [CrossRef]
20. Fang, C.Q.; Yu, R.E.; Liu, S.L.; Li, Y. Nanomaterials applied in asphalt modification: A review. *J. Mater. Sci. Technol.* **2013**, *29*, 589–594. [CrossRef]
21. Hamedi, G.H.; Najad, F.M.; Oveisi, K. Investigating the effects of using nanomaterials on moisture damage of HMA. *Road Mater. Pavement Des.* **2015**, *16*, 536–552. [CrossRef]
22. Akbari, A.; Modarres, A. Effect of clay and lime nano-additives on the freeze-thaw durability of hot mix asphalt. *Road Mater. Pavement Des.* **2017**, *18*, 646–669. [CrossRef]
23. Gong, Y.F.; Bi, H.P.; Tian, Z.H.; Tan, G.J. Pavement performance investigation of nano-TiO_2/$CaCO_3$ and basalt fiber composite modified asphalt mixture under freeze-thaw cycles. *Appl. Sci.* **2018**, *8*, 2581. [CrossRef]
24. Guo, W.; Guo, X.D.; Sun, M.Z.; Dai, W.T. Evaluation of the durability and the property of an asphalt concrete with nano hydrophobic silane silica in spring-thawing season. *Appl. Sci.* **2018**, *8*, 1475. [CrossRef]
25. Guo, W.; Guo, X.D.; Chang, M.Y.; Dai, W.T. Evaluating the effect of hydrophobic nanosilica on the viscoelasticity property of asphalt and asphalt mixture. *Materials* **2018**, *11*, 2328. [CrossRef] [PubMed]
26. Wang, H.P.; Lu, G.Y.; Feng, S.Y.; Wen, X.B.; Yang, J. Characterization of bitumen modified with pyrolytic carbon black from scrap tires. *Appl. Sci.* **2019**, *11*, 1631. [CrossRef]
27. Zhang, P.; Guo, Q.L.; Tao, J.L.; Ma, D.H.; Wang, Y.D. Aging mechanism of a diatomite-modified asphalt binder using Fourier-transform infrared (FTIR) spectroscopy analysis. *Materials* **2019**, *12*, 988. [CrossRef]

© 2019 by the authors. Licensee MDPI, Basel, Switzerland. This article is an open access article distributed under the terms and conditions of the Creative Commons Attribution (CC BY) license (http://creativecommons.org/licenses/by/4.0/).

Article

Assessing the Effect of Nano Hydrophobic Silane Silica on Aggregate-Bitumen Interface Bond Strength in the Spring-Thaw Season

Wei Guo [1], Xuedong Guo [1], Jilu Li [1], Yingsong Li [1], Mingzhi Sun [2] and Wenting Dai [1,*]

[1] School of Transportation, Jilin University, Changchun 130022, China; guowei17@mails.jlu.edu.cn (W.G.); guoxd@jlu.edu.cn (X.G.); jilu16@mails.jlu.edu.cn (J.L.); ysli16@mails.jlu.edu.com (Y.L.)
[2] Research Institute of Highway, Ministry of Transport of China, Beijing 100088, China; mzsun14@mails.jlu.edu.cn
* Correspondence: daiwt@jlu.edu.cn; Tel.: +86-130-8681-4298

Received: 14 May 2019; Accepted: 8 June 2019; Published: 12 June 2019

Featured Application: This paper proposed an aggregate-bitumen interface bond strength test to evaluate the effect of nano hydrophobic silane silica (NHSS) on aggregate-bitumen interface bond strength in the spring-thaw season. The results proved that the addition of NHSS could increase the aggregate-bitumen interface shear strength under any working conditions. Furthermore, the moisture damage model of aggregate-bitumen interface shear strength of NHSS modified asphalt was established based on a research method combining numerical calculations and laboratory tests, which provides suggestions for pavement construction in seasonally frozen regions.

Abstract: In the asphalt–aggregate system, the aggregate-bitumen interface cohesive and adhesive bond determine the mechanical properties of asphalt pavement. The presence of moisture leading to adhesive failure at the binder-aggregate interface and/or cohesive failure within the binder or binder-filler mastic is the main mechanisms of moisture damage in the spring-thaw season. In order to evaluate the effect of nano hydrophobic silane silica (NHSS) on aggregate-bitumen interface bond strength in the spring-thaw season, an aggregate-bitumen interface bond strength test was proposed to quantify the interface bond strength of base asphalt and NHSS modified asphalt. Then, the effect of temperature, freeze-thawing cycles and moisture on aggregate-bitumen interface shear strength of base asphalt and NHSS modified asphalt was also discussed. The results illustrated that the shear failure dominated the aggregate-bitumen interface bonding failure in the spring-thaw season, and temperature and moisture had a significant effect on interface shear strength of modified and unmodified asphalt. Moreover, the addition of NHSS could increase the aggregate-bitumen interface shear strength under any working conditions. Furthermore, the moisture damage model of aggregate-bitumen interface shear strength of base asphalt (BA) and NHSS modified asphalt was established based on a research method combining numerical calculations and laboratory tests.

Keywords: seasonally frozen region; spring-thaw season; nano hydrophobic silane silica; aggregate-bitumen interface; bond strength

1. Introduction

Asphalt mixture has been widely used in the world as the main construction material for pavement because of the advantages including good mechanical strength, smooth ride, low noise, convenient construction and maintenance [1–7]. However, the increase of traffic and the damage from external environmental factors leads to all kinds of pavement diseases, which shorten the service life and increase the maintenance costs [8–12].

Among various pavement diseases, moisture damage is one of the main diseases of early failure of asphalt pavement as it accelerates or causes some typical pavement diseases such as rutting raveling and road surface settlement cracking [13–18]. Especially in seasonally frozen regions, moisture damage is a very serious disease of asphalt pavement caused by the freeze-thaw cycle and dynamic pore water pressure. In the asphalt–aggregate system, the aggregate-bitumen interface cohesive and adhesive bond determine the mechanical properties of asphalt pavement. The loss of cohesion (strength) and stiffness of the asphalt film, and the failure of the adhesive bond between aggregate and asphalt in conjunction with the degradation or fracture of the aggregate were identified as the main mechanisms of moisture damage in the spring-thawing season [19–21].

In recent years, many researchers around the world made great efforts into exploring the moisture damage mechanism at the aggregate-bitumen interface through numerical calculations or laboratory tests [22–24]. Zhang et al. evaluated the moisture sensitivity of aggregate-bitumen bonds with moisture absorption, tensile strength, and failure surface examination. The results showed that the linear relationship between retained tensile strength and the square root of moisture uptake suggests that the water absorption process controls the degradation of the aggregate-bitumen bond [25]. Cho et al. focused on suggesting general mechanisms to explain asphalt-aggregate bond behavior in the Dynamic Shear Rheometer (DSR) moisture damage test. The results indicated that shear-thickening and thixotropy can explain the reversible behavior, the increasing dropping trend in wet conditions in the mechanism of the colloidal system [26]. Caro et al. evaluated the influence of material properties and loading conditions on the response of asphalt mixtures subjected to a moisture environment. The results suggested that the diffusion coefficient of the asphalt matrix and aggregates, and the aggregate-matrix interface bond strength, have the most influence on the moisture susceptibility of the materials [27]. Moraes et al. investigated the feasibility of the newly developed bitumen bond strength (BBS) test for moisture damage characterization. The results indicated that modification and moisture exposure time highly affect the bond strength of asphalt-aggregate systems [28].

In order to solve these problems and extend the service life of asphalt pavements, a lot of research has been conducted on new pavement materials. The prevention approaches of moisture damage mainly focus on increasing aggregate roughness and modified asphalt [29–31]. Among various prevention approaches, it is generally believed that modification of asphalt binders has achieved positive significance over the last few decades [32]. Gorkem et al. determined the effect of additives, such as hydrated lime as well as elastomeric (SBS) and plastomeric (EVA) polymer modified bitumen (PMB) on the stripping potential and moisture susceptibility characteristics of hot mix asphalt. The results indicated that samples prepared with SBS PMB exhibited more resistance to water damage compared to samples prepared with EVA PMB [33]. Palit et al. showed that crumb-rubber modified mixes exhibited lower temperature susceptibility and greater resistance to moisture damage compared to normal mixes [34]. Goh et al. showed that the addition of nanoclay and carbon microfiber would improve a mixture's moisture susceptibility performance or decrease the moisture damage potential in most cases [35].

At present, the modification of asphalt with nanomaterials has gained attention. Nano-material refers to a material with the size of less than 100 nm in at least one dimension. Due to the very small size and huge surface area, the properties of nanomaterials are much different from the normal-sized materials. The addition of these nano-sized particles to another material may overcome the monolithic limitations, and asphalt binder is no exception [36]. In our previous study, the influence of nano hydrophobic silane silica (NHSS) on the property of the asphalt binder and asphalt mixture in the spring-thaw season was discussed systematically. The results showed that the incorporation of NHSS is the most cost-effective technique for mitigating freeze-thaw cycle damage and extending the service life of asphalt pavement in the spring-thaw season [37]. In order to further understand the enhancement mechanism of NHSS in the spring-thaw season, the effect of NHSS on aggregate-bitumen interface bond strength in the spring-thaw season was systematically assessed in this study. Within this framework, an aggregate-bitumen interface bond strength test was proposed to quantify the interface bond strength of base asphalt and NHSS modified asphalt. Then, the effect of temperature, freeze-thawing cycles and moisture on aggregate-bitumen interface shear strength of base asphalt and NHSS modified asphalt was discussed. Moreover, the moisture damage model of aggregate-bitumen interface shear strength of base asphalt and NHSS modified asphalt was established, based on a research method combining numerical calculations and laboratory tests.

2. Materials and Sample Preparation

2.1. Asphalt

The base asphalt used in this paper is AH-90 asphalt named 'Pan Jin' base asphalt. The test parameters of the based asphalt used in this study are summarized in Table 1.

Table 1. Technical parameters of base asphalt.

Technical Parameters	Penetration (0.1 mm)			25 °C Ductility (cm)	Softening Point (°C)	Flash Point (°C)	Solubility (%)	Density (g·cm^{-3})
	15 °C	25 °C	30 °C					
Test results	25.3	86.9	142.4	>130	44.6	340	99.9	1.003
Test procedure	GB/T0606-2011			GB/T0605-2011	GB/T0606-2011	GB/T0611-2011	GB/T0607-2011	GB/T0603-2011

2.2. Nano Hydrophobic Silane Silica

Nano-silica is an important asphalt modifier with superior adhesion, tear resistance and heat aging resistance. However, the nano-silica exhibits hydrophilicity due to the polysilane inside the nano-silica and the reactive silanol groups present on the outer surface. The hydroxyl groups on the surface can be classified into three types: isolated hydroxyl group, continuous hydroxyl group, and twin hydroxyl group. Isolated hydroxyl group is a free hydroxyl group that does not participate in the reaction; the continuous hydroxyl group is formed by two hydrogen groups that generate hydrogen bonds and associate with each other; two hydroxyl groups attached to the same Si are called twin hydroxyl groups. The isolated hydroxyl group and continuous hydroxyl group have no hydrogen bond. The reinforcing effect of nanosilica as a modifier mainly comes from the active structure (-Si-OH) on the surface of the particles, and it is easy to bond the surrounding ions. The surface of nanosilica is uniformly distributed with a layer of silanol and siloxane group with strong water absorption. The water molecules can be physically covered on the surface of the particles or chemically bonded to the hydroxyl groups on the Si atoms. Therefore, nanosilica shows a strong affinity for water. In order to reduce the number of surface silanol structures and change the surface functional groups of the primary nanosilica particles, the surface modifier is used to modify the nanosilica to realize the transition from hydrophilic to hydrophobic. There are many surface treatment agents for nanosilica, and silane coupling agent surface treatment is the most common application method. The nano hydrophobic silane silica is obtained by grafting silane coupling agent onto the surface of nanosilica to carry out surface modification. The surface modification method of nano hydrophobic silane silica is detailed as follows: the X group in the silane coupling agent first undergoes hydrolysis by contacting with moisture, and then forms a temporary oligomer by dehydration condensation. The hydroxyl groups on the surface of nanosilica can react with the oligomeric structure to generate hydrogen bonds, then the nanosilica and the oligomer continue to undergo condensation and dehydration reaction by heating, drying, etc., and finally the silane coupling agent is successfully grafted onto the surface of the nanosilica by a covalent bond, as is shown in Figure 1. After surface modification, the number of silanol structures on the surface of the nanosilica is reduced, the structure of the surface functional groups and the atomic layer of the nanosilica particles is changed. Thereby, the surface properties such as physicochemical adsorption of nanosilica are changed, the surface free energy of the nanoparticle and the phenomenon of agglomeration between particles is reduced. The initial decomposition temperature of nano hydrophobic silane silica obtained by grafting the silane coupling agent on the surface of the silica is about 250 °C, and the loss on ignition (1000 °C, 2 h) is 1.5~2.0, which ensure the workability and effectiveness of nano hydrophobic silane silica in the asphalt pavement construction.

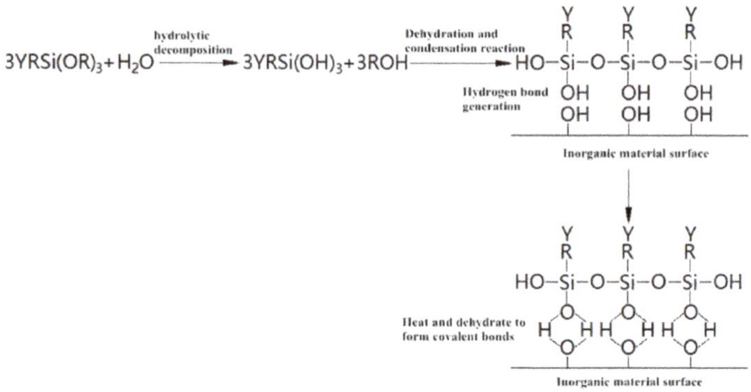

Figure 1. Grafting process of silane coupling agent on nanosilica.

The nano material was obtained from Changtai Micronano Chemical Co., Ltd. (Shouguang, Shandong province, China). The technical properties of nano hydrophobic silane silica material provided by Changtai Weina Chemical Co., Ltd (Shouguang City, China). are presented in Table 2.

Table 2. Technical parameters of nano hydrophobic silane silica.

Technical Parameters	Water Characteristics	BET (m^2/g)	Average Particle Size (nm)	PH Value	SiO$_2$ Content (%)
Test results	Hydrophobia	125 ± 20	12	5.0–8.0	≥99.8
Standard values	——	130 ± 30	≤20	3.7–6.5	≥99.8
Test procedure	GB/T20020	GB/T20020	GB/T20020	GB/T20020	GB/T20020

2.3. Asphalt Sample Preparation

In the previous study, the incorporation method of nano hydrophobic silane silica was discussed in detail [38]. A high shear mixer with the speed of 2000 rpm was used for incorporating the nano hydrophobic silane silica into the base asphalt. Mixing percentages of nano hydrophobic silane silica was 3 wt% of the base asphalt and mixing temperature was kept at 140 °C. The mixing time was about 60 min to ensure homogeneous blending. Base asphalt and nano hydrophobic silane silica modified asphalt were denoted by BA and NHSSMA. The technical parameters of the NHSSMA used in this study are summarized in Table 3.

Table 3. Technical parameters of nano hydrophobic silane silica modified asphalt (NHSSMA).

Technical Parameters	Penetration			10 °C Ductility	Softening Point	PI Value	135 °C Rotational Viscosity
	15 °C	25 °C	30 °C				
Units		0.1 mm		mm	°C	-	mpa
Test results	20.8	64.1	115	233	48.1	−1.357	912.1

3. Aggregate-Bitumen Interface Bond Strength Test

3.1. Test Equipment

The role of asphalt binder in asphalt mixtures is to coat the stones and polymerize the stones together. Therefore, the adhesion theory has long been applied to evaluate the bond strength of hot mix asphalt mixture (HMA). In the adhesion theory, the bond between aggregate and asphalt binder can be ideally considered as an adhesive joint. If more specific, it can be considered as a butt joint. In asphalt mixture system, the asphalt binder is an adherent and the aggregate is considered to be the bonded portion. The destruction of the system occurred within the binder or binder-filler mastic called cohesive failure, which occurred at the binder-aggregate interface considered as adhesive failure, and two failure modes are shown in Figure 2.

Figure 2. Adhesive failure and cohesive failure at the aggregate-bitumen interface.

In this paper, cohesive failure and adhesive failure are collectively referred to as bond failure. In order to measure the aggregate-bitumen interface bond strength of BA and NHSSMA, an aggregate-bitumen interface bond strength tester was designed independently, as show in Figure 3.

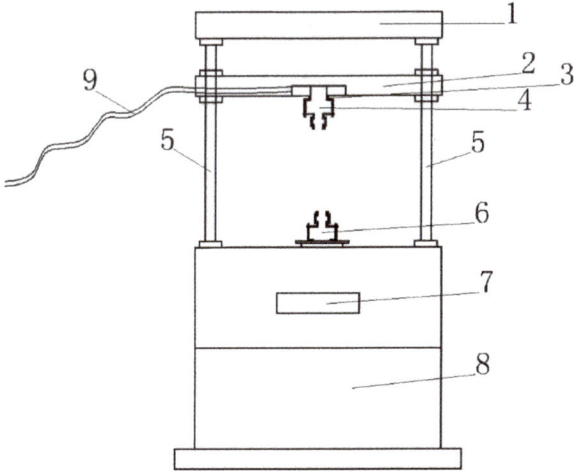

Figure 3. Aggregate-bitumen interface bond strength tester.

The aggregate-bitumen interface bond strength tester is mainly composed of nine main components, which are (1) upper separator, (2) lower separator, (3) hinged bar chain, (4) upper grip finger, (5) screw rod, (6) lower grip finger, (7) asynchronous motor, (8) pedestal and (9) data transmission line. The screw rod (5) of the tester can be driven by the asynchronous motor (7) to shear or stretch the test sample fixed between the upper grip finger (4) and lower grip finger (6). The test sample is champed into the upper grip finger (4) and lower grip finger (6), computer can set the shear (tensile) rate, and the shear or tensile failure load is collected through the data line (9). The shear (tensile) rate was selected to be 1 mm/min. The aggregate-bitumen interface shear (tensile) strength can be calculated as following.

$$SS = \frac{F_s}{S_{sc}}, TS = \frac{F_T}{S_{Tc}},$$

where F_s and F_T is the maximum shear and tensile failure load, S_{sc} and S_{Tc} is the shear and tensile failure area, and SS and TS is the aggregate-bitumen interface shear and tensile strength. The upper grip finger (4) is fixed on the lower separator (2) through the hinged bar chain (3), so that there is only a vertical upward tensile (shear) force during the tensile (shear) process, and no bending moment occurs. The self-developed aggregate-bitumen interface bond strength tester overcomes the disadvantage that the bending moment is easy to generate during the shear or tensile process in the conventional test equipment.

3.2. Test Sample Preparation

In the aggregate-bitumen interface bond strength test, the test mold is a vital part that needs to be designed independently. In order to accurately measure the bond strength and the bond area, the aggregate-bitumen interface must keep flat. In the paper, the appropriate size of granite stones was selected from stone pit, and several square test molds with a size of 40 × 40 × 10 mm was manufactured through large water-cooled stone cutting machine. The size of each test mold after cutting was accurately measured, the unqualified test mold was screened out, and the test mold is shown in Figure 4.

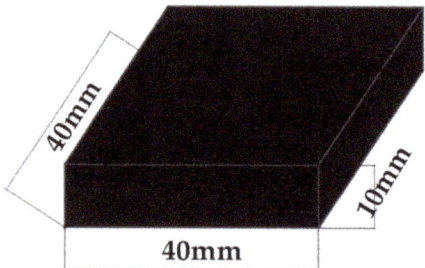

Figure 4. Aggregate-bitumen interface bond strength test mold.

The test sample was prepared by applying the quantitative asphalt binder to the surface of one test mold, flattening with a preheated blade, and then adhering it to the surface of another test mold. The shear sample and tensile sample are shown in Figure 5.

In order to simulate the process of moisture diffusion into the aggregate-bitumen interface, the test sample was placed in a water bath. The test sample placed in the water bath is shown in Figure 6.

As shown in Figure 6, the asphalt binder is not submerged below the water surface. This prevents moisture from entering the aggregate-bitumen interface from the edge of the aggregate-bitumen interface or directly through the asphalt binder, ensuring that moisture can only diffuse from the stone to the aggregate-bitumen interface. The test sample immersed in the water is to evaluate the influence of moisture on the aggregate-bitumen interface bond strength in the spring-thaw season. In addition to the test sample in Figure 6, the water bath (12), the U-shaped heating tube (11) and temperature sensor (10) were also included.

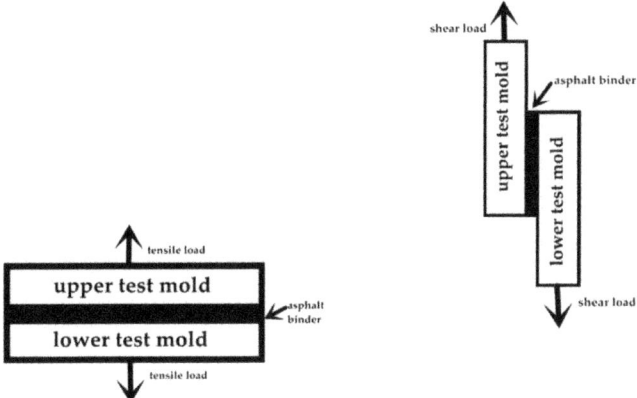

Figure 5. Aggregate-bitumen interface bond strength test tensile sample and shear sample.

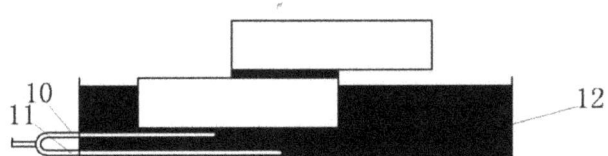

Figure 6. Test shear sample placed in the water bath.

3.3. Test Sample Cleaning

After each aggregate-bitumen interface bond strength test, a layer of asphalt binder remained on the surface of the test mold. Due to the large number of tests, the test mold needed to be cleaned and reused. According to experience, there are two ways to remove surface residual asphalt: (1) high temperature heating cleaning method; (2) chemical reagent cleaning method. Since the high temperature heating cleaning method has high requirements on equipment, the cleaning process is complicated, and the cleaning effect is the same as that of the chemical reagent cleaning method, and repeated high-temperature heat treatment (approximately 500 °C) may adversely affect the physical structure of the test mold. Therefore, the chemical reagent cleaning method was adopted to clean the test mold after test. The specific cleaning process is as follows.

(1) The test mold with residual asphalt was frozen for about 0.5 h.
(2) The test mold was removed from the cryostat and the residual asphalt on the surface of the test mold gently scraped with a plastic sheet.
(3) The surface of the test mold was rinsed with trichloroethylene until no residual bitumen.
(4) The surface of the test mold was rinsed with a small amount of acetone solution to accelerate the drying process of the test mold.
(5) The rinsed test mold was placed in an oven at 100 °C for 12 h.

3.4. Test Procedure

The test equipment, test sample preparation, and test sample cleaning of the aggregate-bitumen interface bond strength test were introduced in the previous section. The details of aggregate-bitumen interface bond strength test are described below.

(1) The test mold was placed in 100 °C oven for 15 min.
(2) BA and NHSSMA were placed in an oven at 135 °C and heated to a flowing state.
(3) The quantitative asphalt binder was applied to the surface of one test mold, flat with a preheated blade, and then adhered to the surface of another test mold.
(4) The prepared test sample was placed in a water bath to simulate the process of moisture diffusion into the aggregate-bitumen interface. The immersion height of each test sample was consistent, and the immersion height was one-half the thickness of the test mold.
(5) The test sample was tested, and the data was obtained by setting the loading rate and the test temperature through a computer. The collected data was the maximum failure load and failure area. In order to ensure the validity and repeatability of the data, the number of parallel tests in each group of experiments was 4–6. The numerical values appearing in the article are the average of the data that meet the error requirements, and the coefficient of variation of all test data was within 10%.
(6) Finally, the test mold was cleaned and we the next set of tests were carried out.

4. Results and Discussion

4.1. Contribution of Shear Strength and Tensile Strength to Aggregate-Bitumen Interface Bond Strength

The load applied to the asphalt binder can be decomposed into tensile and shear loads. In order to study the bond strength of aggregate-bitumen interface under different load effects in the spring-thaw season, the aggregate-bitumen interface shear and tensile strength test for BA and NHSSMA were carried out in −20~20 °C based on the spring temperature. The tensile strength and shear strength of aggregate-bitumen interface for BA and NHSSMA at different temperature are given in Figure 7. It is can be seen from this figure that the aggregate-bitumen interface bond strength was greatly affected following the addition of modifier. From Figure 7, it was observed that addition of nano hydrophobic silane silica improved the shear strength of aggregate-bitumen interface at the temperature above −10 °C, and the increase range is 16.8%~47.1%. The reason for the improvement in shear strength values of nano hydrophobic silane silica modified binder sample is that nano hydrophobic silane silica contribute to the viscosity of asphalt. When the temperature is high, NHSSMA can exert the advantage of high-viscosity property. Therefore, the aggregate-bitumen interface shear strength of NHSSMA is higher than that of BA. However, the asphalt exhibits more elastic characteristics at low temperature. NHSSMA becomes more brittle, so the aggregate-bitumen interface shear strength of BA is slightly larger than that of NHSSMA at −20 °C. It also can be seen from Figure 7 that the aggregate-bitumen interface tensile strength of NHSSMA is slightly larger than that of BA when the temperature is above 0 °C.

From Figure 7, it was also observed that the aggregate-bitumen interface tensile strength was significantly higher than shear strength when the temperature is above −10 °C, but the aggregate-bitumen interface tensile strength was lower than shear strength at −20 °C. Thus, it can be concluded that the aggregate-bitumen interface shear failure is more likely to occur in the pavement when the temperature is above −10 °C, and the aggregate-bitumen interface is prone to tensile failure when the temperature is below −15 °C. This conclusion is also consistent with the fact that the pavement is prone to rutting at high temperatures and cracking is likely to occur at low temperatures. Since the shear failure dominates the aggregate-bitumen interface bonding failure in the spring temperature range, the following study focused on the aggregate-bitumen interface shear failure. In addition, it is interesting to note that the tensile strength of BA and NHSSMA has a peak at −10 °C, which may be related to the glass transition temperature of the asphalt. The glass temperature of asphalt is between −10 and −20 °C. Asphalt will change from glassy state to a viscoelastic state with increasing temperature. When the test temperature is higher than the glass temperature of the asphalt, the asphalt exhibits stronger viscoelasticity, and the adhesion between the asphalt and aggregate increases, resulting in an increase in tensile strength. As the temperature continues to increase, the viscosity of the asphalt gradually decreases, and the tensile strength gradually decreases.

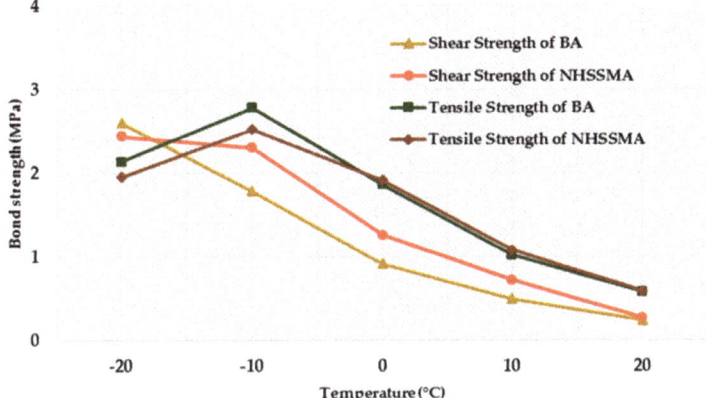

Figure 7. Tensile strength and shear strength of aggregate-bitumen interface for base asphalt (BA) and NHSS modified asphalt (NHSSMA) at different temperature.

4.2. Effect of Temperature on Aggregate-Bitumen Interface Shear Strength

Asphalt is a typical temperature sensitive material, and temperature is the most important factor affecting the aggregate-bitumen shear strength. The aggregate-bitumen interface shear strengths of modified and unmodified asphalt are given in Figure 8. As seen in Figure 8, for modified and unmodified asphalt, the aggregate-bitumen interface shear strength decreased with the increase of temperature, this indicated that, with the increase of temperature, the asphalt became soft and its resistance to interface shear failure declined. Moreover, a higher slope of the fit curve means that a higher temperature sensitivity of the asphalt is observed. As can be witnessed, the temperature sensitivity of nano hydrophobic silane silica modified asphalt was lower than that of base asphalt, which is consistent with the conclusions of the conventional physical properties test in the published article.

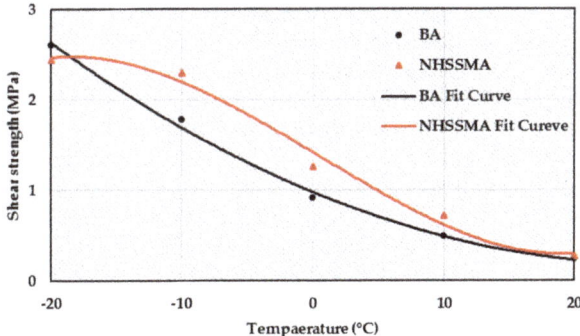

Figure 8. Aggregate-bitumen interface shear strength of BA and NHSSMA at different temperatures.

In the test temperature range of −20~20 °C, the change trend of aggregate-bitumen interface shear strength of NHSSMA with temperature is generally consistent with that of BA. Moreover, the aggregate-bitumen interface shear strength of nano hydrophobic silane silica modified asphalt was always greater than base asphalt when the temperature was above −10 °C, which is because incorporation of nano hydrophobic silane silica can improve the viscosity and resistance to interface shear failure of asphalt. Asphalt tends to be in a glassy state at −20 °C and exhibits more brittleness than viscosity. At the test temperature of −20 °C, the incorporation of nano hydrophobic silane silica not only fails to improve the viscosity of the asphalt but also destroys the integrity and uniformity of asphalt, thereby causing the aggregate-bitumen interface shear strength of NHSSMA to be lower than that of BA at −20 °C.

4.3. Effect of Freeze-Thaw Cycles on Aggregate-Bitumen Interface Shear Strength

In the seasonally frozen regions, freeze-thaw (F-T) damage is the main pavement disease, causing a variety of poor conditions in bitumen pavement, such as cracks, pits, potholes, and slush. In order to evaluate the effect of freeze-thaw cycles on aggregate-bitumen interface shear strength of NHSSMA, damaged BA and NHSSMA after 10, 20, and 30 F-T cycles were collected for an aggregate-bitumen interface shear strength test. The details of a freeze-thaw cycle test is described below.

First, neat and modified asphalt binders were heated to a fluid state and poured into a fixed-size plate to ensure the dimensions of asphalt binder samples was approximately 6 × 250 × 250 mm. The purpose of this was to make the thickness of the prepared asphalt binder samples as close as possible to the thickness of the asphalt film in the actual pavement.

Then, the samples were submerged in a container containing water, then the container with specimens were placed in the precision temp-enclosure at −15 °C and frozen for 10 h.

Finally, the base asphalt and modified asphalt samples were soaked in water at 15 °C for 16 h through adjusting the temperature controller.

As described above, a complete freeze-thaw aging cycle was completed. Then, after 10, 20, and 30 freeze-thaw aging cycles, damaged samples were collected for aggregate-bitumen interface shear strength test to evaluate the effect of freeze-thaw cycles on aggregate-bitumen interface shear strength of NHSSMA. Moreover, another set of experiments was designed to consider the effect of snow-melting agent on asphalt in the spring-thaw season. A 20% concentration of $CaCl_2$ solution was selected for the F-T cycle. The aggregate-bitumen interface shear strength of modified and unmodified asphalt with freeze-thaw cycles under different conditions are given in Figure 9.

As shown in Figure 9a, 10 °C aggregate-bitumen interface shear strength of BA increases with the increase of F-T cycles, and then decreases slightly. After 30 F-T cycles, 10 °C aggregate-bitumen interface shear strength of BA decreased by 18.9% to 0.4 MPa. The 10 °C aggregate-bitumen interface shear strength of NHSSMA decreased with the increase of F-T cycles. After 30 F-T cycles, 10 °C aggregate-bitumen interface shear strength of NHSSMA decreased by 24.6% to 0.546 MPa. It can be seen that 10 °C aggregate-bitumen interface shear strength decay rate of NHSSMA is greater than BA under the F-T cycles, but 10 °C aggregate-bitumen interface shear strength of NHSSMA is still higher than that of BA at different F-T cycles. This indicates that the incorporation of NHSS increases the 10 °C aggregate-bitumen interface shear strength of asphalt under the F-T cycles.

It can be concluded from Figure 9b that the snow-melting agent has a greater influence on 10 °C aggregate-bitumen interface shear strength, and the decay rate under a snow-melting agent is higher than that under clear water. After 30 F-T cycles, 10 °C aggregate-bitumen interface shear strength of BA and NHSSMA decreased by 20.3% and 43.4%, respectively. However, whether it is a clear water or $CaCl_2$ solution, 10 °C aggregate-bitumen interface shear strength of NHSSMA is greater than that of BA. This shows that NHSS can improve the aggregate-bitumen interface shear strength of asphalt in the spring-thaw season.

As shown in Figure 9c,d, after 30 F-T cycles, −10 °C aggregate-bitumen interface shear strength of BA and NHSSMA decreased by 4.8% and 21.8%, respectively. After 30 F-T cycles, −10 °C aggregate-bitumen interface shear strength of BA and NHSSMA with snow-melting agent decreased by 10% and 31.3%, respectively. The variation of aggregate-bitumen interface shear strength with F-T cycles at −10 °C is basically the same as that of 10 °C.

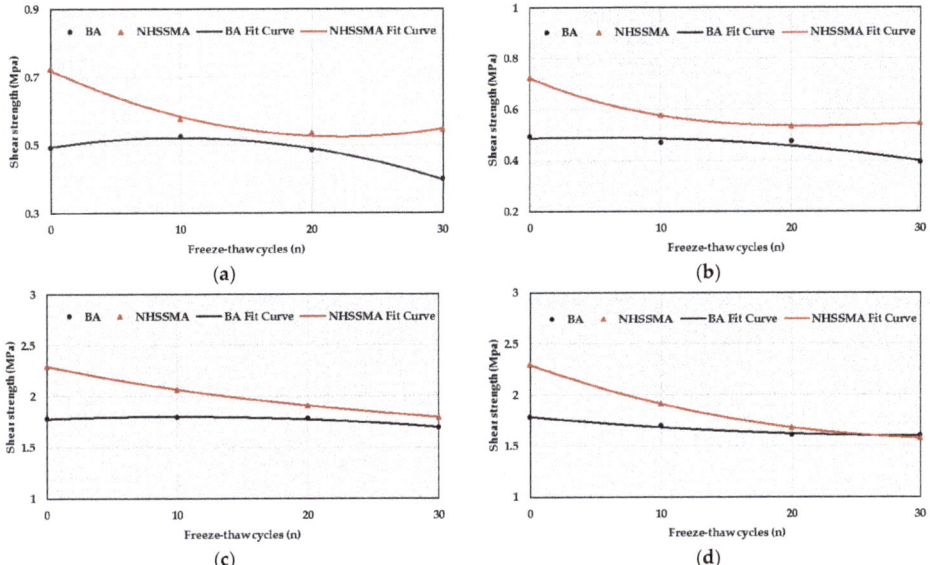

Figure 9. The changes of aggregate-bitumen interface shear strength of BA and NHSSMA with freeze-thaw cycles under different conditions. (**a**) The 10 °C aggregate-bitumen interface shear strength of BA and NHSSMA under different freeze-thaw cycles. (**b**) The 10 °C aggregate-bitumen interface shear strength of BA and NHSSMA with snow-melting agent under different freeze-thaw cycles. (**c**) The −10 °C aggregate-bitumen interface shear strength of BA and NHSSMA under different freeze-thaw cycles. (**d**) The −10 °C aggregate-bitumen interface shear strength of BA and NHSSMA with snow-melting agent under different freeze-thaw cycles.

In summary, the aggregate-bitumen interface shear strength of modified and unmodified asphalt decreases with the increase of F-T cycle, which is due to the aging of asphalt caused by the F-T cycle. The snow-melting agent has a greater influence on aggregate-bitumen interface shear strength, and the decay rate under the snow-melting agent is higher than that under clear water. This may be due to the chemical reaction of chloride ions in the snow-melting agent with the asphalt, further aggravating the aging of the asphalt. Whether it is a clear water or $CaCl_2$ solution, the aggregate-bitumen interface shear strength of NHSSMA is greater than that of BA. This indicates that NHSS can improve the aggregate-bitumen interface shear strength of asphalt in the spring-thaw season. Moreover, the data also show that the aggregate-bitumen interface shear strength of modified and unmodified asphalt is less affected by F-T cycles at low temperatures relative to high temperatures. This also explains that pavement moisture damage mainly occurs in warm springs rather than cold winters.

4.4. Effect of Moisture on Aggregate-Bitumen Interface Shear Strength

In the spring-thaw season, after the asphalt film breaks, the aggregate-bitumen interface will be exposed to the moisture environment and the penetration speed of the moisture to the aggregate-bitumen interface will increase exponentially compared with the diffusion speed through the asphalt film. In order to evaluate the variation of the shear strength of the aggregate-bitumen interface that is directly exposed to the moisture environment, the dry test shear sample was completely immersed in a water bath at 10 °C for 4, 8, 24, and 48 h and then taken out for testing. The test results include the average aggregate-bitumen interface shear strength and coefficient of variation (CV). The formula for CV is as follows.

$$CV = \frac{\sigma}{\mu},$$

where σ is data standard deviation and μ is data average. The test results are given in Figure 10. It can be seen from Figure 10 that 10 °C aggregate-bitumen interface shear strength of BA and NHSSMA gradually decrease with increasing immersion time, and the shear strength decreased significantly in the early stage of water immersion. After 4 h of immersion, 10 °C aggregate-bitumen interface shear strength of NHSSMA and BA are greatly reduced, with a decrease of 43.1% and 48.3%, respectively. After that, 10 °C aggregate-bitumen interface shear strength of modified and unmodified asphalt decreases slowly as the immersion time increases. After 48 h of immersion, 10 °C aggregate-bitumen interface shear strength of NHSSMA and BA are reduced by 51% and 55.4%, respectively. This indicates that when the asphalt film has a large range of rupture, the moisture can quickly penetrate into the aggregate-bitumen interface, which has a significant effect on the aggregate-bitumen interface shear strength. Moreover, whether it is strength value or the descend range, NHSSMA is better than BA at different immersion times, which indicates that the addition of nano hydrophobic silane silica improves the ability of asphalt binder to resist moisture damage.

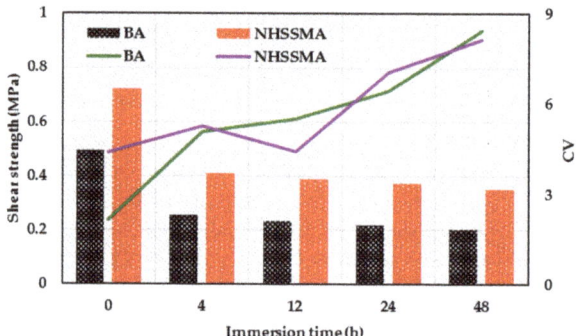

Figure 10. The 10 °C aggregate-bitumen interface shear strength of BA and NHSSMA at different immersion times.

As seen in Figure 10, CV of test results for modified and unmodified asphalt gradually increase with the immersion time, and CV of test results for the two types of asphalt binders is smallest when the immersion time is zero. This indicates that the aggregate-bitumen interface shear strength test can obtain more accurate data in the dry environment, and the longer the immersion time, the greater the discreteness of the test data. CV of all test results measured in this paper is within 8.4%, which proves that the aggregate-bitumen interface shear strength test has good repeatability.

4.5. Establishment of Moisture Damage Model of Aggregate-Bitumen Interface Shear Strength

The aggregate-bitumen interface bond strength plays an important role in the pavement performance of bituminous pavement. Generally speaking, the stone is more hydrophilic compared with asphalt. In the spring-thaw season, after the moisture penetrates into the aggregate-bitumen interface, the moisture tends to wet more stone surface, which causes the bond failure between stone and asphalt, forcing the asphalt film to peel off from the stone surface. In order to evaluate the influence of interfacial moisture content on aggregate-bitumen interface shear strength of NHSSMA, a research method combining numerical calculations and laboratory tests was proposed. First, the relation between aggregate-bitumen interface shear strength and immersion time was established through laboratory tests. Secondly, the relationship between immersion time and interfacial moisture content was determined by Abaqus finite element software. Finally, the relationship between the interfacial moisture content and aggregate-bitumen interface shear strength was established by the intermediate indicator of the immersion time.

The laboratory test refers to the aggregate-bitumen interface bond strength test of Section 3, the test shear sample was immersed in water at 10 °C and the immersion height was half the thickness of the lower test mold, approximately 5 mm. After 4, 24, 48, 96, and 144 h of water immersion, the damaged test shear sample was collected for shear test immediately. When the test temperature of shear test was −10 °C, the damaged test shear sample needed to be packaged with plastic wrap and placed in a temperature control box at −10 °C for 2 h. The shear test was carried out after the temperature of the test shear sample was stabilized at −10 °C. The test results are shown in Figure 11.

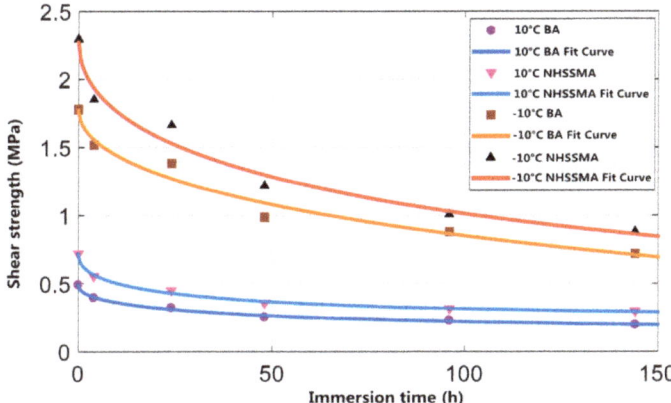

Figure 11. Aggregate-bitumen interface shear strength of BA and NHSSMA at different immersion times.

It can be seen from Figure 11 that the short-dated immersion can greatly reduce the aggregate-bitumen interface shear strength. After 48 h of water immersion, the interfacial moisture content tends to be stable, the 10 °C aggregate-bitumen interface shear strength of modified and unmodified asphalt are reduced by 48.3% and 50.3%, and the −10 °C aggregate-bitumen interface shear strength of modified and unmodified asphalt are reduced by 44.5% and 47%, respectively. Thereafter, as the immersion time increases, the aggregate-bitumen interface shear strength of modified and unmodified asphalt enters a stage of steady decline. When the immersion time reached 144 h, 10 °C aggregate-bitumen interface shear strength of BA and NHSA asphalt are reduced by 59.9% and 59.3%, and −10 °C aggregate-bitumen interface shear strength of modified and unmodified asphalt is reduced by 59.7% and 61.6%, respectively. The 10 or −10 °C aggregate-bitumen interface shear strength of nano hydrophobic silane silica modified asphalt is higher than that of base asphalt by nearly 80% at any immersion time, and the modification effect is remarkable.

Through the above tests, the relationship between aggregate-bitumen interface shear strength of modified and unmodified asphalt and immersion time was obtained. In order to further explore the effect of interfacial moisture content on aggregate-bitumen interface shear strength, it is necessary to quantify interfacial moisture content.

In this paper, the Mass Diffusion module of Abaqus finite element software was applied for interfacial moisture content calculation. Abaqus/Standard provides transient and steady-state models for calculating the diffusion of one material to another. The control equation used in the calculation was an extended form of Fick diffusion law, which can be used to calculate the diffusion of a non-uniform solubility material in the base material. The control equations and constitutive relations used in the calculation are as follows.

(1) Control equation

The calculation of the percolation diffusion issue obeys the mass conservation law of the diffusion phase.

$$\int_V \frac{dc}{dt} dV + \int_S n \cdot J dS = 0,$$

where c is mass concentration of diffusion material, V is any volume, S is surface of V, n is the out-of-plane normal direction of S, J is the diffusion flux of the diffusion phase, and t is the time. Apply the divergence theorem to get the following formula.

$$\int_V \left(\frac{dc}{dt} + \frac{\partial}{\partial x} \cdot J \right) dV = 0.$$

Due to the arbitrariness of the volume, the following formula can be obtained.

$$\int_V \delta\phi \left(\frac{dc}{dt} + \frac{\partial}{\partial x} \cdot J \right) dV = 0,$$

where $\delta\phi$ is an arbitrary continuous scalar field. The above formula can be extended to the following formula.

$$\int_V \left[\delta\phi \left(\frac{dc}{dt} \right) + \frac{\partial}{\partial x} \cdot (\delta\phi J) - J \cdot \frac{\partial \delta\phi}{\partial x} \right] dV = 0,$$

where ϕ is the standard concentration, $\varphi = c/s$, s is the solubility of the diffusion material in the base material. The following formula is obtained by applying the divergence theorem again.

$$\int_V \left[\delta\phi \left(\frac{dc}{dt} \right) - \frac{\partial \delta\phi}{\partial x} \cdot J \right] dV + \int_S \delta\phi n \cdot J dS = 0.$$

(2) Constitutive relationship

Diffusion is assumed to chemical potential gradient drive, and the basic constitutive relationship is as shown in the following equation.

$$J = -sD \cdot \left[\frac{\partial \phi}{\partial x} + k_s \frac{\partial}{\partial x} (ln(\theta - \theta^z)) + k_p \frac{\partial p}{\partial x} \right],$$

where, $D(c, \theta, f)$ is the diffusion coefficient; $s(\theta, f)$ is solubility; $k_s(c, \theta, f)$ is the Sorel effect factor, related to the temperature gradient and affects the diffusion; θ is the temperature; θ^z is all zero degrees; $k_p(c, \theta, f)$ is the compressive stress factor and is related to the equivalent compressive stress gradient p, which affects the diffusion; f is any predefined field variable.

The size of simulated model is consistent with the size of the actual shear specimen, which is composed of two $40 \times 40 \times 10$ mm cubic granite stones. The main parameter is the moisture diffusion coefficient of the rock material (D_{rock}), and D_{rock} is 0.6 mm^2/h according to the granite stone. The boundary condition is that the water immersion is half of the height of the lower block, so the water concentration of all surfaces below the 1/2 height of the block is 1. The collected data is the interfacial moisture content at different immersion times in the calculation, and the total immersion time is 144 h. All conditions of the calculation are consistent with the actual situation. The interfacial moisture content is calculated as follows.

$$\phi = \frac{C_m}{C_m^{max}},$$

where ϕ is the interfacial moisture content, C_m is the interfacial moisture concentration, C_m^{max} is the extreme limit of interfacial moisture concentration. If $\phi = 1$, it means that the moisture at the interface has reached a state of saturation. Figure 12 shows the moisture content of shear specimen at 4 h of immersion.

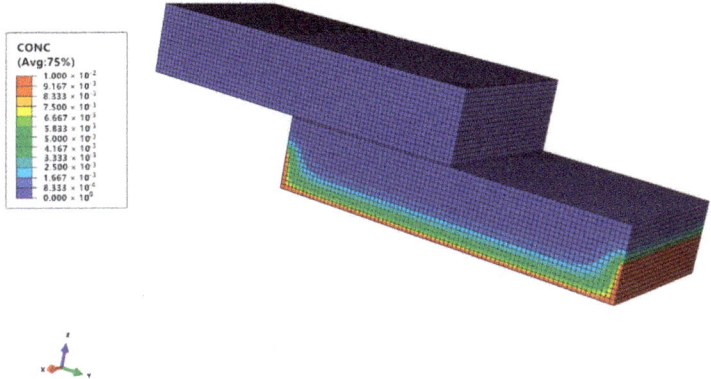

Figure 12. Moisture content of shear specimen at 4 h immersion time.

Figure 13 shows fitting curve of interfacial moisture content with immersion time. It can be seen from Figure 13 that interfacial moisture content increases rapidly at the early stage of immersion, and can reach 60% of the extreme limit of interfacial moisture concentration after 50 h of immersion. After 100 h of immersion, the interfacial moisture content tends to be saturated, and the growth rate is obviously slowed down.

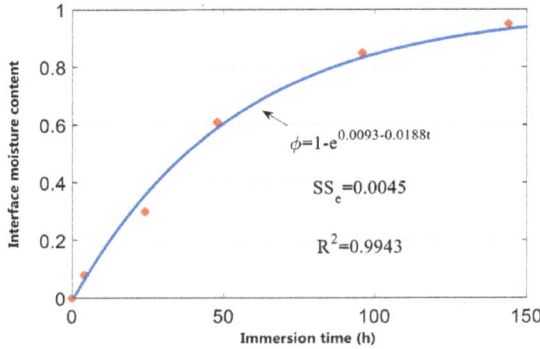

Figure 13. Fitting curve of interfacial moisture content with immersion time.

According to the test results and numerical calculation results, the relationship between aggregate-bitumen interface shear strength and interfacial moisture content can be obtained, as shown in Table 4.

Table 4. Fitting results of the relationship between aggregate-bitumen interface shear strength and interfacial moisture content.

Asphalt Category	Test Temperature	Regression Equation	Equation Parameters			SS_e	R^2
			a	b	c		
BA	10 °C		0.193	0.276	0.721	0.0001	0.99
BA	−10 °C	$SS = e^{a-b\sqrt{\phi}} - c$	2.806	0.066	14.71	0.03	0.96
NHSSMA	10 °C		−0.221	0.764	0.084	0.0001	0.99
NHSSMA	−10 °C		3.897	0.029	46.94	0.023	0.98

SS_e is residual sum of squares, R^2 is determination coefficient.

According to Table 4, it can be clearly found that the fitting result is better at higher temperatures, indicating that acquired data is relatively stable at higher temperatures and has better regularity, and the dispersion of data is slightly larger at lower temperatures. In order to establish aggregate-bitumen interface shear strength damage model of modified and unmodified asphalt binder, a residual shear strength index (RSS) is proposed. The residual shear strength is calculated as follows.

$$\text{RSS} = \frac{SS_n}{SS_0} \times 100,$$

where RSS is residual aggregate-bitumen interface shear strength, SS_0 is the aggregate-bitumen interface shear strength when interfacial moisture content is 0, SS_n is the aggregate-bitumen interface shear strength at different interfacial moisture content. Figure 14 shows fitting curve of 10 °C residual aggregate-bitumen interface shear strength of BA with interfacial moisture content.

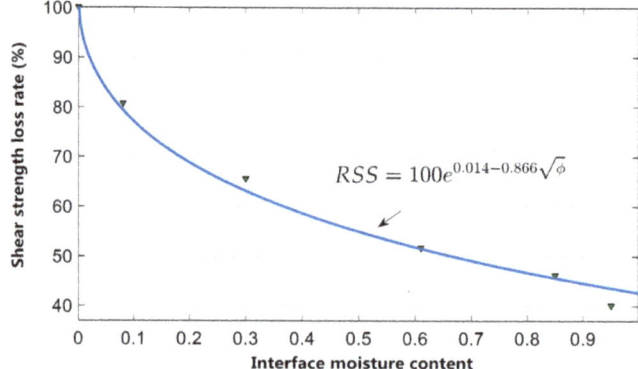

Figure 14. Fitting curve of 10 °C residual aggregate-bitumen interface shear strength of BA with interfacial moisture content.

The regression equation of damage model and the corresponding value of each model is shown in Table 5. The moisture damage model of aggregate-bitumen interface shear strength is effective, because all R^2 of the regression equation are above 0.9. In order to compare the model parameters and analyze physical meaning, the model parameter a is defined as shear strength damage degree, and the value of b represents the aggregate-bitumen interface shear strength damage rate. It can be clearly seen from Table 5 that the damage degree of NHSSMA is larger than BA in the test temperature range, which indicates that the modified asphalt is more sensitive to interfacial moisture, especially at lower temperature. Interestingly, NHSSMA is more sensitive to interfacial moisture, but it can still provide greater aggregate-bitumen interface shear strength than BA at any interfacial moisture content. This may be due to the fact that the incorporation of the nano hydrophobic silane silica does not change the moisture-sensitive properties of aggregate-bitumen interface, but it can improve the viscosity of asphalt, thus the incorporation of nano hydrophobic silane silica improves the bond strength between asphalt and aggregate in a moisture environment.

Table 5. Fitting results of moisture damage model of aggregate-bitumen interface shear strength.

Asphalt Category	Test Temperature	Regression Equation	Equation Parameters		SS_e	R^2
			a	b		
BA	10 °C		0.014	0.866	21.85	0.99
BA	−10 °C	$RSS = e^{a-b\sqrt{\phi}}$	0.039	0.791	189.2	0.93
NHSSMA	10 °C		0.0008	0.899	3.035	0.99
NHSSMA	−10 °C		0.025	0.866	118.8	0.95

It can be seen from Table 5 that the damage factor b of modified and unmodified asphalt at low temperature is greater than that at high temperature, which indicates that the damage rate of aggregate-bitumen interface shear strength is faster at higher temperature. Aggregate-bitumen interface shear strength of BA at −10 °C is about 3.6~4.3 times that at of BA at 10 °C, and aggregate-bitumen interface shear strength of NHSSMA at −10 °C is about 3~3.7 times that at of NHSSMA at 10 °C. Thus, the aggregate-bitumen interface bond failure is most likely to occur in the spring-thaw season.

5. Conclusions

The effect of hydrophobic nano-silica on aggregate-bitumen interface bond strength in spring-thaw season was assed.

In this paper, the effect of nano hydrophobic silane silica on the interface bond strength in spring-thaw season was systematically investigated according to self-designed aggregate-bitumen interface bond strength test. The results indicated that the addition of nano hydrophobic silane silica improves the ability of aggregate-bitumen interface to resist moisture damage in the spring-thaw season. The main conclusions are as follows:

1. The aggregate-bitumen interface bond strength test results illustrated that the shear failure dominated the aggregate-bitumen interface bonding failure in the spring-thaw season.
2. The aggregate-bitumen interface shear strength test showed that temperature and moisture had a significant effect on interface shear strength of modified and unmodified asphalt. Moreover, the addition of NHSS could increase the aggregate-bitumen interface shear strength under any working conditions.
3. Based on a research method combining numerical calculations and laboratory tests, the moisture damage model of aggregate-bitumen interface shear strength of BA and NHSSMA was established. Model parameter a and parameter b could be used to evaluate the moisture damage degree and moisture damage rate of aggregate-bitumen interface shear strength, respectively.

Author Contributions: Data curation, W.G., Y.L. and M.S.; Formal analysis, W.G.; Funding acquisition, X.G.; Investigation, Y.L. and M.S.; Methodology, W.G.; Project administration, X.G. and W.D.; Software, W.G.; Writing—original draft, W.G., J.L. and M.S.; Writing—review and editing, X.G. and W.D.

Funding: This research was funded by the National Nature Science Foundation of China (NSFC) (Grant No. 51178204).

Conflicts of Interest: The authors declare no conflict of interest.

References

1. Zhang, D.; Chen, M.Z.; Wu, S.P.; Riara, M.; Wan, J.M.; Li, Y.Y. Thermal and rheological performance of asphalt binders modified with expanded graphite/polyethylene glycol composite phase change material (EP-CPCM). *Constr. Build. Mater.* **2019**, *194*, 83–91. [CrossRef]
2. El-Hakim, M.; Tighe, S.L. Impact of freeze-thaw cycles on mechanical properties of asphalt mixes. *Transp. Res. Rec.* **2014**, *2444*, 20–27. [CrossRef]
3. Porto, M.; Caputo, P.; Loise, V.; Eskandarsefat, S.; Teltayev, B.; Rossi, C.O. Bitumen and bitumen modification: A review on latest advances. *Appl. Sci.* **2019**, *9*, 742. [CrossRef]
4. Rossi, C.O.; Teltayev, B.; Angelico, R. Adhesion promoters in bituminous road materials: A review. *Appl. Sci.* **2017**, *7*, 524. [CrossRef]
5. Ji, J.; Yao, H.; Liu, L.H.; Suo, Z.; Zhai, P.; Yang, X.; You, Z.P. Adhesion evaluation of asphalt-aggregate interface using surface free energy method. *Appl. Sci.* **2017**, *7*, 156. [CrossRef]
6. Wang, W.T.; Wang, L.B.; Xiong, H.C.; Luo, R. A review and perspective for research on moisture damage in asphalt pavement induced by dynamic pore water pressure. *Constr. Build. Mater.* **2019**, *204*, 631–642. [CrossRef]
7. Mozaffari, S.; Tchoukov, P.; Atias, J.; Czarnecki, J.; Nazemifard, N. Effect of Asphaltene Aggregation on Rheological Properties of Diluted Athabasca Bitumen. *Energy Fuels* **2015**, *29*, 5595–5599. [CrossRef]

8. Luo, R.; Huang, T.T.; Zhang, D.R.; Lytton, R.L. Water vapor diffusion in asphalt mixtures under different relative humidity differentials. *Constr. Build. Mater.* **2017**, *136*, 126–138. [CrossRef]
9. Molenaar, A.A.A.; Hagos, E.T.; van de Ven, M.F.C. Effects of aging on the mechanical characteristics of bituminous binders in PAC. *J. Mater. Civ. Eng.* **2010**, *22*, 779–787. [CrossRef]
10. Mozaffari, S.; Tchoukova, P.; Mozaffari, A.; Atias, J.; Czarnecki, J.; Nazemifard, N. Capillary driven flow in nanochannels—Application to heavy oil rheology studies. *Colloids Surf. A.* **2017**, *513*, 178–187. [CrossRef]
11. Nian, T.F.; Li, P.; Wei, X.Y.; Wang, P.H.; Li, H.S.; Guo, R. The effect of freeze-thaw cycles on durability properties of SBS-modified bitumen. *Constr. Build. Mater.* **2018**, *187*, 77–88. [CrossRef]
12. Darjani, S.; Koplik, J.; Pauchard, V. Extracting the equation of state of lattice gases from random sequential adsorption simulations by means of the Gibbs adsorption isotherm. *Phys. Rev. E* **2017**, *96*, 052803. [CrossRef] [PubMed]
13. Airey, G.D.; Collop, A.C.; Zoorob, S.E.; Elliott, R.C. The influence of aggregate, filler and bitumen on asphalt mixture moisture damage. *Constr. Build. Mater.* **2008**, *22*, 2015–2024. [CrossRef]
14. Wang, H.; Li, E.Q.; Xu, G.J. Molecular dynamics simulation of asphalt-aggregate interface adhesion strength with moisture effect. *Int. J. Pavement Eng.* **2017**, *18*, 414–423. [CrossRef]
15. Cucalon, L.G.; Bhasin, A.; Kassem, E.; Little, D.; Herbert, B.E.; Masad, E. Physicochemical characterization of binder-aggregate adhesion varying with temperature and moisture. *J. Transp. Eng. Part B Pavement.* **2017**, *143*, 04017007. [CrossRef]
16. Berdahl, P.; Akbari, H.; Levinson, R. Weathering of roofing materials—An overview. *Constr. Build. Mater.* **2008**, *22*, 423–433. [CrossRef]
17. Liu, F.; Darjani, S.; Akhmetkhanova, N.; Maldarelli, C.; Banerjee, S.; Pauchard, V. Mixture Effect on the Dilatation Rheology of Asphaltenes-Laden Interfaces. *Langmuir* **2017**, *33*, 1927–1942. [CrossRef]
18. Lin, Y.M.; Hu, C.C.; Adhikari, S.; Wu, C.H.; Yu, M. Evaluation of Waste Express Bag as a Novel Bitumen Modifier. *Appl. Sci.* **2019**, *9*, 1242. [CrossRef]
19. Kok, B.V.; Yilmaz, M. The effects of using lime and styrene-butadiene-styrene on moisture sensitivity resistance of hot mix asphalt. *Constr. Build. Mater.* **2009**, *23*, 1999–2006. [CrossRef]
20. Sengoz, B.; Agar, E. Effect of asphalt film thickness on the moisture sensitivity characteristics of hot-mix asphalt. *Build. Sci.* **2007**, *42*, 3621–3628. [CrossRef]
21. Yang, X.L.; Shen, A.Q.; Guo, Y.C.; Zhou, S.B.; He, T.Q. Deterioration mechanism of interface transition zone of concrete pavement under fatigue load and freeze-thaw coupling in cold climatic areas. *Constr. Build. Mater.* **2018**, *160*, 588–597. [CrossRef]
22. Caro, S.; Masad, E.; Bhasin, A.; Little, D. Moisture susceptibility of asphalt mixtures, Part 1: Mechanisms. *Int. J. Pavement Eng.* **2008**, *9*, 81–98. [CrossRef]
23. Preeda, C.; Hussain, U.B. Effect of moisture on the cohesion of asphalt mastics and bonding with surface of aggregates. *Road Mater. Pavement Des.* **2018**, *19*, 741–753. [CrossRef]
24. Apeagyei, A.K.; Grenfell, J.R.A.; Airey, G.D. Observation of reversible moisture damage in asphalt mixtures. *Constr. Build. Mater.* **2014**, *60*, 73–80. [CrossRef]
25. Zhang, J.Z.; Airey, G.D.; Grenfell, J.; Apeagyei, A.K. Moisture damage evaluation of aggregate-bitumen bonds with the respect of moisture absorption, tensile strength and failure surface. *Road Mater. Pavement Des.* **2017**, *18*, 833–848. [CrossRef]
26. Cho, D.W.; Kim, K. The mechanisms of moisture damage in asphalt pavement by applying chemistry aspects. *KSCE J. Civil Eng.* **2010**, *14*, 333–341. [CrossRef]
27. Caro, S.; Masad, E.; Bhasin, A.; Little, D. Micromechanical modeling of the influence of material properties on moisture-induced damage in asphalt mixtures. *Constr. Build. Mater.* **2010**, *24*, 1184–1192. [CrossRef]
28. Moraes, R.; Velasquez, R.; Bahia, H.U. Measuring the effect of moisture on asphalt-aggregate bond with the bitumen bond strength test. *Transp. Res. Rec.* **2011**, *2209*, 70–81. [CrossRef]
29. Yao, H.; Dai, Q.L.; You, Z.P. Chemo-physical analysis and molecular dynamics (MD) simulation of moisture susceptibility of nano hydrated lime modified asphalt mixtures. *Constr. Build. Mater.* **2015**, *101*, 536–547. [CrossRef]
30. Tarefder, R.A.; Zaman, A.M. Nanoscale evaluation of moisture damage in polymer modified asphalts. *J. Mater. Civ. Eng.* **2010**, *22*, 714–725. [CrossRef]

31. Rossi, C.O.; Ashimova, S.; Calandra, P.; De Santo, M.P.; Angelico, R. Mechanical resilience of modified bitumen at different cooling rates: A rheological and atomic force microscopy investigation. *Appl. Sci.* **2017**, *7*, 779. [CrossRef]
32. Yusoff, N.I.M.; Breem, A.A.S.; Alattug, H.N.M.; Hamim, A.; Ahmad, J. The effects of moisture susceptibility and ageing conditions on nano-silica/polymer-modified asphalt mixtures. *Constr. Build. Mater.* **2014**, *72*, 139–147. [CrossRef]
33. Gorkem, C.; Sengoz, B. Predicting stripping and moisture induced damage of asphalt concrete prepared with polymer modified bitumen and hydrated lime. *Constr. Build. Mater.* **2009**, *23*, 2227–2236. [CrossRef]
34. Palit, S.K.; Reddy, K.S.; Pandey, B.B. Laboratory evaluation of crumb rubber modified asphalt mixes. *J. Mater. Civ. Eng.* **2004**, *16*, 45–53. [CrossRef]
35. Goh, S.W.; Akin, M.; You, Z.P.; Shi, X.M. Effect of deicing solutions on the tensile strength of micro- or nano-modified asphalt mixture. *Constr. Build. Mater.* **2011**, *25*, 195–200. [CrossRef]
36. Mozaffari, S.; Li, W.H.; Thompson, C.; Ivanov, S.; Seifert, S.; Lee, B.; Kovarik, L.; Karim, A.M. Colloidal nanoparticle size control: Experimental and kinetic modeling investigation of the ligand metal binding role in controlling the nucleation and growth kinetics. *Nanoscale* **2017**, *9*, 13772–13785. [CrossRef]
37. Guo, W.; Guo, X.D.; Sun, M.Z.; Dai, W.T. Evaluation of the durability and the property of an asphalt concrete with nano hydrophobic silane silica in spring-thawing season. *Appl. Sci.* **2018**, *8*, 1475. [CrossRef]
38. Guo, W.; Guo, X.D.; Chang, M.Y.; Dai, W.T. Evaluating the effect of hydrophobic nanosilica on the viscoelasticity property of asphalt and asphalt mixture. *Materials* **2018**, *11*, 2328. [CrossRef] [PubMed]

© 2019 by the authors. Licensee MDPI, Basel, Switzerland. This article is an open access article distributed under the terms and conditions of the Creative Commons Attribution (CC BY) license (http://creativecommons.org/licenses/by/4.0/).

Article

Experimental Investigation into the Structural and Functional Performance of Graphene Nano-Platelet (GNP)-Doped Asphalt

Murryam Hafeez [1,*], Naveed Ahmad [1], Mumtaz Ahmed Kamal [1], Javaria Rafi [1], Muhammad Faizan ul Haq [1], Jamal [2], Syed Bilal Ahmed Zaidi [1] and Muhammad Ali Nasir [3]

1. Department of Civil Engineering, University of Engineering and Technology Taxila, Taxila 47080, Pakistan; n.ahmad@uettaxila.edu.pk (N.A.); drmakamal@yahoo.com (M.A.K.); javariarafi@outlook.com (J.R.); faizan.ul@students.uettaxila.edu.pk (M.F.u.H); bilal.zaidi@uettaxila.edu.pk (S.B.A.Z.)
2. Department of Civil Engineering, Royal Melbourne Institute of Technology University, Victoria 3053, Australia; jamal.naasir209@gmail.com
3. Department of Mechanical Engineering, University of Engineering and Technology Taxila, Taxila 47080, Pakistan; ali.nasir@uettaxila.edu.pk
* Correspondence: murryamhafeez77@outlook.com; Tel.: +92-3222189592

Received: 28 December 2018; Accepted: 13 February 2019; Published: 17 February 2019

Abstract: With the increase in the demand for bitumen, it has become essential for pavement engineers to ensure that construction of sustainable pavements occurs. For a complete analysis of the pavement, both its structural and functional performances are considered. In this study, a novel material (i.e., Graphene Nano-Platelets (GNPs)) has been used to enhance both of the types of pavements' performances. Two percentages of GNPs (i.e., 2% and 4% by the weight of the binder) were used for the modification of asphalt binder in order to achieve the desired Performance Grade. GNPs were homogeneously dispersed in the asphalt binder, which was validated by Scanning Electron Microscope (SEM) images and a Hot Storage Stability Test. To analyze the structural performance of the GNPs-doped asphalt, its rheology, resistance to permanent deformation, resistance to moisture damage, and bitumen-aggregate adhesive bond strength were studied. For the analysis of the functional performance, the skid resistance and polishing effect were studied using a British Pendulum Skid Resistance Tester. The results showed that GNPs improved not only the rutting resistance of the pavement but also its durability. The high surface area of GNPs increases the pavement's bonding strength and makes the asphalt binder stiffer. GNPs also provide nano-texture to the pavement, which enhances its skid resistance. Thus, we can recommend GNPs as an all-around modifier that could improve not only the structural performance but also the functional performance of asphalt pavements.

Keywords: Graphene nano-platelets (GNPs); asphalt; Scanning Electron Microscope (SEM); structural performance; functional performance

1. Introduction

Bitumen is considered to be an essential component of roadways and its demand is increasing with each passing day. According to the Asphalt Institute, 87 million tons of bitumen are produced per year around the globe [1]. A major chunk of this production, approximately 85% of the bitumen, is used in the paving industry. With the increase in traffic loads, the need for progress in pavement technology is also increasing. The early failure of pavement calls for its reconstruction, which results in an increased demand for bitumen. In order to conserve the resources, it is necessary to ensure the construction of sustainable pavements.

In Pakistan, two commonly observed highway failures are rutting and moisture damage. The poor mix properties of asphalt and the high temperature greatly contribute to these failures. The temperature cannot be controlled; however, the properties of asphalt can be improved for better temperature resistance. For over 50 years, researchers have been using various asphalt modifiers to achieve the desired material properties. Recently, the use of nanoparticles for asphalt modification has been brought into the hot spot because of their unique properties. Researchers have used various kinds of nanoparticles for enhancing the properties of asphalt, such as Nanosilica, Carbon Nanotubes (CNTs), Carbon Black Nanoparticles (CBNPs), and Graphite Nanoparticles (xGNPs). Graphene Nano-Platelets (GNPs) have commendable mechanical and thermal properties and as a result, their applications can be found in a broad range of fields [2–5]. A high specific surface area (SSA) and shape ratio (diameter/thickness) are responsible for imparting these properties on GNPs [6]. The strong carbon-carbon bond not only contributes to their exceptional strength but also provides chemical and structural stability [7]. The modification of asphalt with GNPs results in enhanced adhesive forces that increase the moisture resistance of asphalt. Asphalt modified by Graphene Nano-Platelets (GNPs) has been found to have improved mechanical and compaction properties when compared to the conventional asphalt [8]. The two challenges that researchers have to face while working with nanoparticles are their high cost and difficulty in homogeneously dispersing them in an asphalt binder. While working with GNPs, researchers overcame both of these challenges, as GNPs are low in cost and it is easier to achieve their homogeneous dispersion in an asphalt binder [8,9].

Researchers use a wet mixing technique to disperse nanoparticles in the binder. Nanoparticles are first dispersed in the solvent using a mechanical stirrer and then are mixed with the binder using a high shear mixer. The commonly observed issue with the wet mixing technique is that, if the solvent does not evaporate completely from the binder, it compromises the properties of the binder [1]. The usage of the solvent and the high shear mixer adds to the extra cost, as well as increases the processing time. On the other hand, it is comparatively easy to disperse GNPs in the asphalt binder, as no solvent or shear mixer is required. This makes the industrial application of GNPs favorable.

In order to evaluate a pavement, a structural and functional analysis of it is carried out. The structural performance is related to the pavement's strength and capacity to carry loads and traffic flow during its service life. The functional performance relates to the roughness of the pavement's surface. The skid resistance is an important parameter of the functional performance when it comes to the safety of the pavement. It is influenced by the micro-texture and the macro-texture of the pavement. The micro-texture affects the skid resistance in the pavement's early life, whereas the macro-texture influences the skid resistance over the service life of the pavement. To study the micro-texture and macro-texture of asphalt, Scanning Electron Microscopy and a British Pendulum Skid Resistance Tester are used. The structural and functional performances are inter-dependent. In a structurally sound pavement, the macro-texture remains intact for a longer period of time, providing skid resistance [10]. Normally, the gradation of aggregates is altered to get the desired skid resistance. In this study, we have worked with GNPs to improve the structural and functional performances of pavement, using a single modifier at the same time.

As the introduction of GNPs into the world of pavements happened quite recently, their impact on the rutting resistance, moisture susceptibility, and skid resistance of the asphalt needs further exploration. This paper not only aims to study the structural performance of asphalt modified by GNPs but to also explore its functional performance.

2. Materials and Methods

2.1. Materials

2.1.1. Graphene Nano-Platelets

Graphene Nano-Platelets are basically sheets of graphene piled up together. A single graphene sheet is a monolayer of carbon atoms. These carbon atoms are tightly packed in a hexagonal arrangement. The stacking, rolling, and wrapping of graphene results in the formation of graphite, CNTs, and fullerenes, respectively. Thus, graphene can be labeled as a building block in the formation of the allotropes of carbon [11].

For this study, GNPs were procured from Advanced Chemical Suppliers (ACS) Materials, Pasadena, CA, USA. These GNPs were prepared using the interlayer cleavage method. Table 1 shows the properties of the procured GNPs. The shape ratio (diameter/thickness) and specific surface area are the two main properties of GNPs that are responsible for imparting the structural strength to GNPs-doped asphalt.

Table 1. The properties of Graphene Nano-Platelets (GNPs).

Appearance	Black/Grey Powder
Diameter	2–7 µm
Thickness	2–10 nm
Specific Surface Area	20–40 m^2/g
Electrical Conductivity	80,000 S/m
Carbon Content	>99%
Apparent Density	0.06–0.09 g/ml
Water Content	<2 wt.%
Residual Impurities	<1 wt.%
Unit Price	$1.9/gram

2.1.2. Bitumen

For this study, 60/70 Penetration Grade bitumen was used. It was procured from the Attock Oil Refinery Limited, Pakistan.

2.1.3. Aggregates

The source of the aggregates used for this research was Margalla Quarry, Punjab, Pakistan. It is a local quarry of limestone. Table 2 shows the qualitative properties of the aggregates obtained from the Margalla Hills [12].

Table 2. The properties of the aggregates procured from Margalla Quarry.

Material Properties	Standard	Results (%)	NHA * Specification Limits
Fractured particles	ASTM D5821	100	90% (min)
Flakiness	BS 812.108	4.75	10% (max)
Elongation	BS 812.109	2.2	10% (max)
Sand equivalent value	ASTM D 2419	75	50% (min)
Los Angeles abrasion	ASTM C 131	23	30% (max)
Water absorption	ASTM C 127	1.02	2% (max)
Soundness (Coarse)	ASTM C 88	7.1	8% (max)
Soundness (Fine)	ASTM C 88	4.7	8% (max)
Uncompacted voids	ASTM C 1252	45.5	45% (min)

* National Highway Authority.

The National Highway Authority Class B (for wearing course), which is specified as a finer gradation, was used for the preparation of the asphalt mixtures. Figure 1 represents the midpoint gradation curve for the National Highway Authority (NHA) Class B.

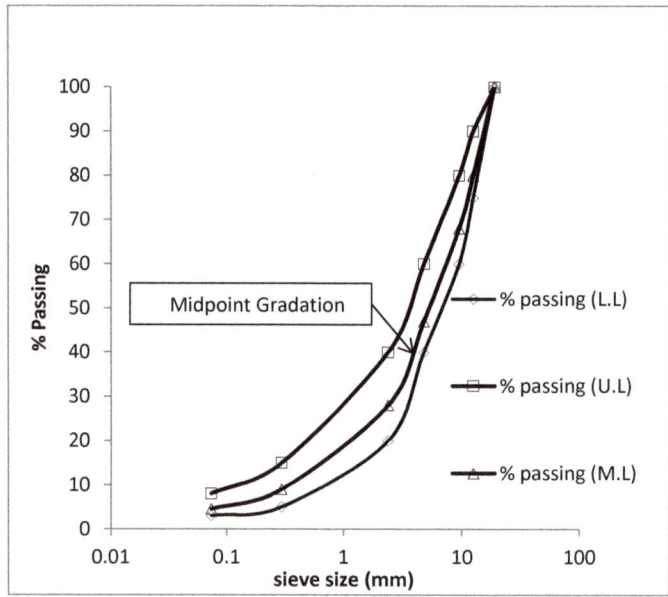

Figure 1. The aggregate gradation curve.

2.2. Preparation of GNP-Modified Asphalt

In Pakistan, the asphalt binder selection is currently based upon penetration grading, whereas pavement engineers around the globe are adopting performance grading (PG). The concept of performance grading is based around the idea that the properties of the asphalt binder should be in accordance with the requirements put forth by the environmental conditions of the area in which it is used. According to the temperature zoning of Pakistan, in 70% of areas the required performance grade of a binder is 70-10 [13,14]. Unfortunately, PG 70-10 binder is not produced in any of the oil refineries in Pakistan [13]. The PG of the binder used in this study was 58-22, which is much softer than the required PG 70-10. The high PG values are more of a concern, as the temperature hardly falls below 0 °C in most of the areas of Pakistan. The aim of the study was to achieve a PG of 70 using GNPs. For this research, two percentages of GNPs were used (2% and 4%) by weight (23% and 46% by volume) of the binder. This choice was made on the basis of the performance grade, as PG 64 was achieved when 2% of GNPs were added to the binder and PG 70 was achieved upon the addition of 4% of GNPs. For performance grading, a Dynamic Shear Rheometer was used. The frequency was set as 10 rad/s and 25 mm geometry was used. The initial temperature was set at 58 °C. The highest temperature, where the value of $G^*/\sin\delta$(kPa) was reduced to 1.0 kPa or less, was termed the high-temperature performance grade (PG) of the binder. The variation in the performance grade of the binder after its modification with GNPs is presented in Figure 2. The PG of the base binder was 58 °C and it failed at 62.7 °C. Upon the addition of 2% of GNPs, the PG increased by one level, i.e., 64 °C. The failure temperature also increased from 62.7 °C to 65.7 °C. A more pronounced change was observed when 4% of GNPs were added to the asphalt binder. PG 70 was achieved after modifying the binder with 4% of GNPs and the failure temperature was 71.5 °C. The trend shows that, with the addition of GNPs to the asphalt binder, its performance grade increases.

Prior to the mixing process, the asphalt binder was heated in an oven for 30 min at 158 ± 5 °C. A glass rod was used to mix the GNPs into the binder for 10 min at 158 ± 5 °C [8]. In order to ensure homogeneous mixing of the GNPs in the asphalt binder, Scanning Electron Microscope (SEM) (Vega3, TESCAN, Czech Republic) images were taken. The Scanning Electron Microscope has been shown in

Figure 3. Figure 4 shows the SEM image of the GNPs, in which the plate-like/flaky morphology of the GNPs can be seen. Figure 5 is the SEM image of the GNPs dispersed in the asphalt binder. In the image, homogeneously dispersed GNPs are visible in the binder.

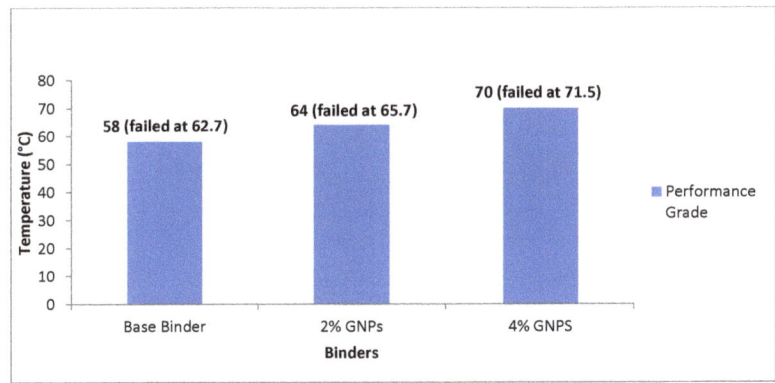

Figure 2. The performance grades (PGs) of the asphalt binders.

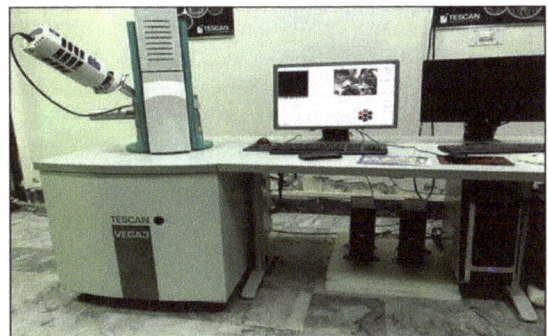

Figure 3. The Scanning Electron Microscope (SEM) in the Mechanical Engineering Department (University of Engineering and Technology, Taxila).

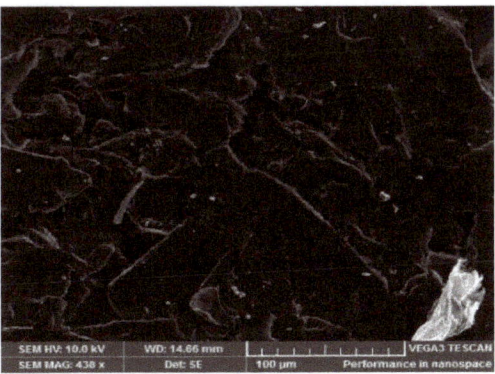

Figure 4. A Scanning Electron Microscope (SEM) image of the Graphene Nano-Platelets (GNPs) (100 μm).

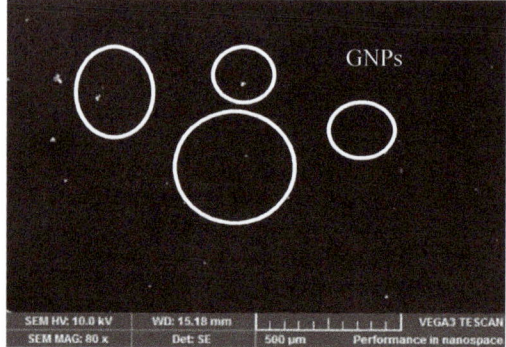

Figure 5. A Scanning Electron Microscope (SEM) image of the Graphene Nano-Platelets (GNPs) in the asphalt binder (500 µm).

According to the General Specifications of the National Highway Authority (NHA) of Pakistan, two aggregate gradations named class "A" and "B" are used for the construction of asphalt pavement wearing courses. Class A is coarser, while class B is finer [15]. For this study, the asphalt mixtures were prepared using the NHA class B aggregate gradation for the wearing course (19.5 mm Nominal Maximum Aggregate Size). To determine the Optimum Binder Content (OBC) of the GNP-modified asphalt, the Marshall Method of Mix Design was adopted. The asphalt mixes were prepared using a base binder and GNP-modified binders. Throughout the design, 5.5% of the air voids were maintained, which is the mean of the specified range of the in-place air voids (3% to 8%) according to the Asphalt Institute Manual Series-2 (MS-2). The OBCs of the three types of mixes are reported in Table 3.

Table 3. The Optimum Binder Content (OBC) of the asphalt mixtures.

Type of Mix	Optimum Binder Content
Base Binder	4.38
2% GNPs	4.49
4% GNPs	4.67

2.3. Storage Stability Test

The storage stability is an integral parameter to consider while working with binders modified with nanoparticles. A Hot Storage Stability Test was carried out to check whether the suspension of the GNPs in the bitumen was stable for storage or not. This test was performed according to BS EN 13399 (2017). During this, 50 g of GNP-modified binder was poured into an aluminum tube. It was kept in the oven at 163 °C for 72 h. After the sample was taken out, it was cut into three sections. The top and bottom sections were heated and poured into rings for a softening point test. For a stable storage sample, the difference in the softening point of the top and bottom sections should be less than 2.2 °C. Although the softening point test is the most common method to determine the storage stability, it is not highly accurate [16]. For accuracy, the performance grades of all of the three sections were also determined by using a Dynamic Shear Rheometer (DSR) to ensure the homogeneous mixing of the GNPs in the binder.

The softening point results for the top and bottom sections of the tube are shown in Table 4.

Table 4. The softening point test results for storage stability.

Softening Point (°C)	Base Binder	2% Graphene Nano-Platelets (GNPs)	4% Graphene Nano-Platelets (GNPs)
Top Section	48.5	49.8	57.4
Bottom Section	49	50.5	57.8
Difference	0.5	0.7	0.4

According to the results, the difference between the softening point of the top and bottom sections of the 2% and 4% GNP-modified binder was less than 2.2 °C. This indicates that the GNP-modified asphalt binder is a stable storage blend [17].

The performance grades of the top and bottom sections of the samples for all of the blends were also determined using a DSR to ensure the homogeneous dispersion of the GNPs in the asphalt binder, as the rheological characterization of the binder is affected by its composition [16]. The performance grades of the binders did not vary between the sections.

2.4. Study of the Structural Performance

2.4.1. Conventional Testing and Rheological Analysis of the Binder

- Conventional Binder Testing

The penetration, softening point, and ductility tests were performed in accordance with ASTM D5-13, ASTM D36-76, and ASTM D113-99, respectively.

- Study of the Rheology of the Binder Using a Dynamic Shear Rheometer

The rheological parameters of the binder, such as the Complex shear modulus (G^*), Phase angle (δ), and Superpave Rutting factor ($G^*/\sin\delta$), were determined using an Anton Paar Dynamic Shear Rheometer (DSR). Two tests were performed in accordance with AASHTO T 315 (High-Temperature Performance Grading and a Frequency Sweep Test).

The range of frequencies used for the frequency sweep test was 10 to 0.1 rad/s. The range of temperatures used was 20 °C to 70 °C. Both of the geometries (8 mm and 25 mm) were used. In this test, oscillatory shear stress is used at a constant strain level to determine the storage and loss modulus at the given range of frequencies and temperatures. The test data obtained were fitted to develop master curves using the sigmoidal function, as described by the mechanistic-empirical pavement design guide (MEPDG). Master curves were constructed at a reference temperature of 50 °C by giving a horizontal shift to the data obtained from each test temperature. A temperature of 50 °C was selected because the main focus of the study was to analyze the high-temperature performance of the binder. Before the Frequency Sweep test, a Strain Sweep test was performed at each testing temperature in strain-controlled mode. The complex shear modulus (G^*) was measured and the percent strain value at which the complex shear modulus was reduced to 95% of its initial value was noted as a threshold for the linear viscoelastic (LVE) region. Based on this, the strain limit for the base binder was kept at 10%, whereas it was 0.45% for the GNP-modified binder.

2.4.2. Permanent Deformation Analysis

- Study of the Rut Resistance

The rut resistance of asphalt was studied at 40 °C and 55 °C using a Cooper Wheel Tracking Machine. The standard followed for this test was BS EN12697-22. The asphalt mix was prepared at 158 ± 5 °C and compacted to form slabs, with the help of a compacting machine, until the target of 5.5% air voids was achieved. Three slabs were prepared for each test condition. The dimensions of the slabs were 305 mm × 305 mm × 50 mm. The load applied by the wheel tracking machine was 700 N. The thickness of the wheel was 50.8 mm and its diameter was 203.2 mm. The speed of the machine was 26.5 rpm. Each slab was subjected to 10,000 loading cycles and the rut depth was measured. The wheel-tracking slope should be expressed in mm per 10^3 load cycles and calculated using Formula (1).

$$WTS_{AIR} = (d_{10000} - d_{5000})/5 \tag{1}$$

where WTS_{AIR} is the wheel-tracking slope (mm/10^3 load cycles) and d_{5000} and d_{10000} are the rut depths (mm) after 5000 load cycles and 10,000 load cycles, respectively.

- Determination of the Dynamic Modulus of Asphalt

The ratio of the peak-to-peak stress to the peak-to-peak strain is termed the Dynamic Modulus. A Dynamic Modulus test was performed using the Servo-Pneumatic Universal Testing Machine NU-14 (Cooper, Ripley, UK). This test is also termed as a viscoelastic test in which sinusoidal loading is applied. The standard followed for this test was in accordance with AASHTO TP 62-8. Cylindrical specimens of asphalt were prepared using a Superpave Gyratory Compactor (Controls, Milan, Italy), with a diameter of 152.4 mm and a height of 172.72 mm. Samples with a diameter of 101.6 mm and a height of 152.4 mm were extracted using a core cutter machine. Three samples were prepared against each test condition. The test was performed at 40 °C and 55 °C. The loading frequencies were 25, 10, 5, 1, 0.5, and 0.1 Hz. The loads applied at 40 °C and 55 °C were 195 kPa and 53 kPa, respectively. The Dynamic Modulus was recorded with the help of Linear Variable Differential Transformers (LVDTs).

2.4.3. Durability Analysis

- Moisture Susceptibility Analysis (Rolling Bottle Test)

A Rolling Bottle test was carried out to study the moisture sensitivity of the asphalt. The test was performed in accordance with BS EN 12697-11:2005. The 6.3 mm to 10 mm of aggregates' fraction sieved was used as per the standard EN 12697-2. The weight of the aggregates was 170 g. The weight of the binder was 8 g. The aggregates were heated and uniformly coated with the binder. 150 g of aggregate coated with binder was placed in the bottle. 400 mL of distilled water was added to the bottle. Two samples were prepared against each binder. All of the bottles were placed in the rolling bottle machine. The speed of the machine was 60 revolutions per minute. The samples were taken out and studied for the percentage of bitumen coverage after 6, 24, 48, and 72 h.

- Bitumen-Aggregate Adhesion Analysis

The Bitumen Bond Strength was studied using a Pneumatic Adhesion Tensile Test Instrument (PATTI). The standard followed for this test was ASTM D 4541. Plates of limestone were prepared by cutting out slabs. The dimensions of the slabs were 381 mm × 152.4 mm × 50.8 mm. The slabs were cleaned for 60 min in an ultrasonic cleaner at 60 °C and then heated in the oven at 150 °C for an hour. Prior to the start of the test, the binder, stubs, and slabs were heated in the oven for 30 min at 75 °C. The type of stubs used for the test was F-4, with a diameter of 12.7 mm. 0.04g of bitumen was placed on the stub and then the stub was placed on the slab. Three samples were prepared against each binder. The samples were tested after 12 h of conditioning at room temperature. The PATTI gives the value of burst pressure at which the stub gets detached from the substrate, which was converted to the Pull-Off Tensile Strength (POTS) by using the following Formula (2).

$$POTS = [(BP * Ag) - C]/Aps \tag{2}$$

where:

Ag = Contact area of the gasket with the reaction plate = 2619.35 mm^2
BP = Burst pressure (MPa)
Aps = Area of the pull stub = 126.64 mm^2
C = Piston Constant = 129.73 g.

2.5. Study of the Functional Performance

2.5.1. Study of the Surface Texture

In order to study the surface texture of aggregates coated with GNP-modified binder, images from a digital camera were captured. To observe it in detail, SEM images were used. The samples were prepared and sputtered before observation under the SEM.

2.5.2. Study of the Skid Resistance and Polishing Effect

A British Pendulum Skid Resistance Tester was used to study the skid resistance of the asphalt. This test was performed in accordance with ASTM E303-93(2013). Before subjecting the asphalt slabs to the Wheel Tracking Machine, their surfaces were tested for their skid resistance. The British Pendulum Skid Resistance Tester gave us a British Pendulum Number (BPN) against each asphalt mix. The higher the BPN, the higher the skid resistance is. After the completion of the Wheel Tracking Test (10,000 loading cycles), the skid resistance was checked again on the path developed due to the continuous movement of the wheel. This was done in order to study the polishing of the asphalt surface due to the movement of the wheel on the asphalt.

3. Results and Discussion

3.1. Conventional Testing Results

The results of the conventional tests performed in the laboratory to study the physical properties of the binders are presented in Table 5.

Table 5. The physical properties of the base and modified binders.

Test	Base Binder	2% Graphene Nano-Platelets (GNPs)	4% Graphene Nano-Platelets (GNPs)
Penetration (0.1 mm)	61	40	32
Softening Point (°C)	48	49.5	57
Ductility (cm)	100	14	13

It was observed that the addition of the GNPs to the binder reduced its penetration value. Reductions of 34% and 48% in the penetration values were recorded after the addition of 2% and 4% of GNPs, respectively. An inverse relationship between the GNP content and binders' penetration value can be seen. An increase in the GNP content led to a decrease in the penetration value, which validated the stiffening of the binder. In the same way, the results of the softening point also depict the stiffening of the binder. The softening point of the binder increased up to 19% when 4% of GNPs were added. The improvement in the physical properties is due to the increase in the bonding strength after the homogeneous dispersion of GNP layers in the binder, which restricts the flow of bitumen and makes it stiffer [18]. The extremely small size and high surface area of GNPs enable them to form a strong bond with bitumen, which leads to a reduction in the penetration value and an increase in the softening point [19]. A massive reduction in the ductility was also recorded in the case of GNP-modified binder, which can be attributed to increased stiffness [20]. The trend in the physical properties of the GNP-modified binder indicates an enhancement in its high-temperature performance.

3.2. Rheology of the Binder

Frequency Sweep Test

Rheological changes were observed in the asphalt binder upon the addition of GNPs. An increase in the complex shear modulus and a decrease in the phase angle values were recorded. Figures 6 and 7 show the Master Curves for the Complex Shear Modulus and Phase Angle, respectively. It is evident in Figure 6 that modification of the binder with GNPs has led to an increase in the complex modulus. The increment in the complex modulus was more pronounced when 4% of GNPs were added in comparison to 2%. Figure 7 shows a decrement in the phase angle, which is more prominent at a low frequency/high temperature. A maximum decrease in the phase angle was observed for the higher weight content of the GNPs. The value of the Superpave rutting factor also increased massively at a low frequency upon modification of the binder. This showed the same trend as the Complex Shear Modulus.

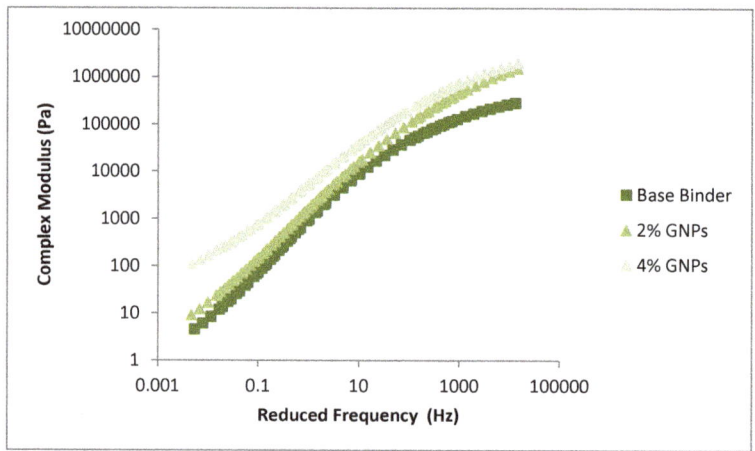

Figure 6. The Master Curve for the Complex Modulus at 50 °C. GNPs = Graphene Nano-Platelets.

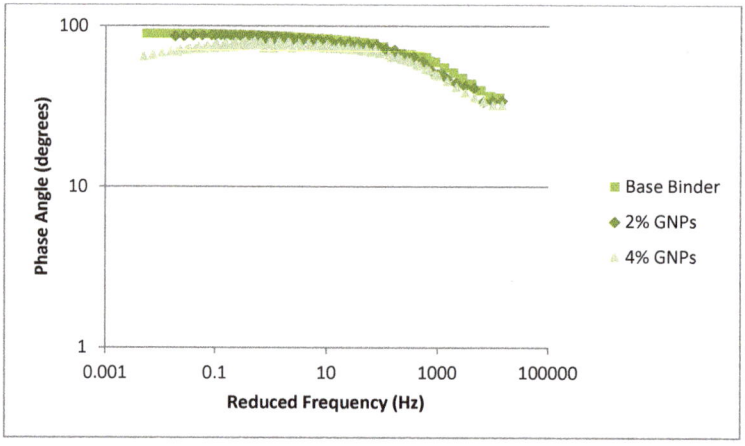

Figure 7. The Master Curve for the Phase Angle at 50 °C. GNPs = Graphene Nano-Platelets.

The reason behind the increase in the complex modulus can be attributed to the high elasticity and the large surface area of the nanoparticles, which gives them a higher affinity to bond with the functional groups of the binder [21]. A cover is created, which helps in preventing the viscous nature of bitumen and causes a delay in the conversion of the elastic behavior to the viscosity at a high temperature [22]. The downward shift of the phase angle can also be attributed to the elasticity of the GNPs. The elastic behavior of the asphalt binder contributes to the rutting resistance. Increasing the GNP content intensifies the increase in the complex modulus and the decrease in the phase angle.

The increase in the Superpave rutting factor can be explained by the fact that the extremely small size of the GNPs leads to an increase in Van der Waals interaction energy between the GNPs and the asphalt binder [23]. This results in a more stable asphalt binder. The enhancement in the Superpave rutting factor due to the addition of the GNPs validates the improvement in the deformation resistance at a high temperature.

3.3. Rutting Resistance

The Wheel Tracking Test was performed at two temperatures (40 °C and 55 °C). Figures 8 and 9 show the relationship between the number of passes and the rut depth for all of the samples. The results depict an enhancement in the rut resistance of the asphalt following the addition of the GNPs. The increment in the rut resistance is more pronounced at 55 °C, where the rut depth decreased from 7.3 mm to 4.7 mm after the addition of 4% of GNPs. At 40 °C, the rut depth decreased from 2.8 mm to 2.59 mm when 2% of GNPs were added, and further decreased to 2.45 mm following the addition of 4% of GNPs. Table 6 shows the results of the Wheel Tracking Slope. It can be seen that, with the increase in the content of the GNPs, the values of the slope decrease at 55 °C and remain constant at 40 °C. A decrease in the slope means that there is an increase in resistance to a permanent deformation [24]. According to the results, the maximum resistance to a permanent deformation can be achieved at 55 °C by adding 4% of GNPs. Due to the high surface area of the GNPs, they are reactive and form a strong bond with bitumen, imparting strength [25]. The GNPs may also act as filler, increasing the structural asphalt and significantly reducing the temperature susceptibility of asphalt [26].

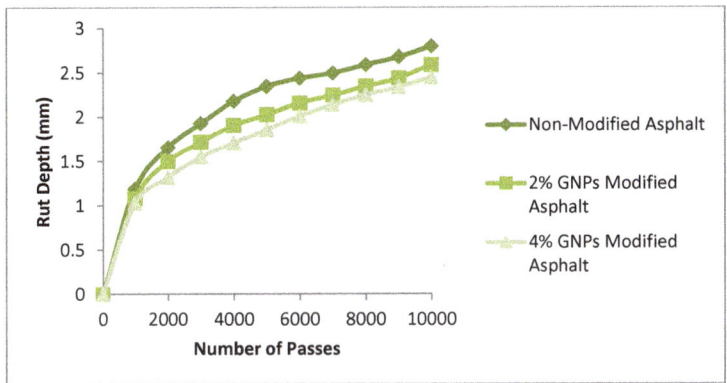

Figure 8. The Wheel Tracking Test results at 40 °C. GNPs = Graphene Nano-Platelets.

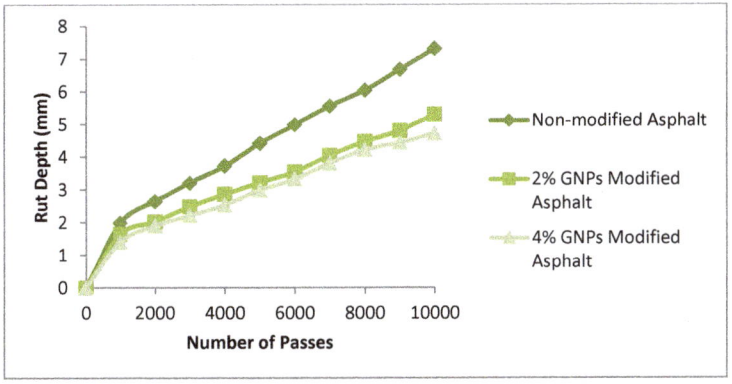

Figure 9. The Wheel Tracking Test results at 55 °C. GNPs = Graphene Nano-Platelets.

Table 6. The Wheel Tracking slope. GNPs = Graphene Nano-Platelets.

	Wheel Tracking Slope (mm/10³ Load Cycles)	
Temperature	40 °C	55 °C
Non-Modified Asphalt	0.09	0.58
2% GNPs Modified Asphalt	0.11	0.42
4% GNPs Modified Asphalt	0.12	0.35

3.4. Dynamic Modulus

The Dynamic Modulus Test was performed using NU-14 at 40 °C and 55 °C. Figures 10 and 11 show the relationship between the frequency and the dynamic modulus for the samples at both of the temperatures. Dynamic Modulus is a direct measure of the rut resistance. It is evident from the results that the value of the Dynamic Modulus is higher for GNP-modified asphalt. Asphalt modified with 4% of GNPs gave the highest values of Dynamic Modulus, regardless of the temperature and loading conditions. The same trend has been observed in the Wheel Tracking Test too. In general, the values of the Dynamic Modulus at 55 °C are lower than the values at 40 °C, which indicates a greater rut susceptibility at a high temperature. It can also be seen that the load response lag phenomenon is quite obvious when the loading frequency increases. This indicates that the Dynamic Modulus or strength of the asphalt mixture increases with an increment in the loading frequency [27]. The well-dispersed GNPs can provide support to asphalt by increasing the interaction and adhesion strengths within it [28]. These effects may lead to an increase in the Dynamic Modulus and resistance to the rutting of GNP-doped asphalt.

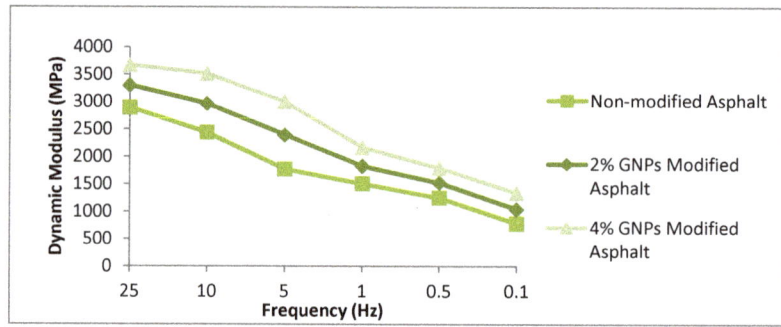

Figure 10. The Dynamic Modulus at 40 °C. GNPs = Graphene Nano-Platelets.

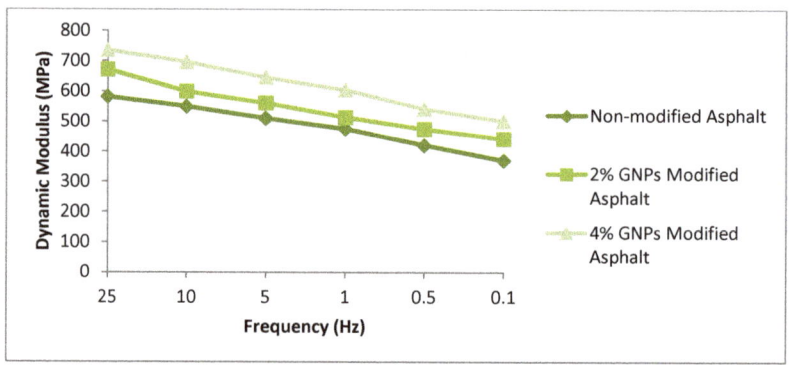

Figure 11. The Dynamic Modulus at 55 °C. GNPs = Graphene Nano-Platelets.

3.5. Moisture Susceptibility

The percentage of bitumen coverage was noted after 6, 24, 48, and 72 h during the rolling bottle test. Figure 12 is the graphical representation of the relationship between the time of rolling and the percentage of bitumen coverage. Figure 13 shows the affinity between the bitumen and aggregate after 72 h of rolling. The rolling bottle test is a measure of the binder's moisture susceptibility. A higher percentage of bitumen coverage indicates a strong adhesive bond between the binder and the aggregate, as well as a high resistance of the binder to moisture damage. It is evident from the results that, after the completion of the test, the percentage of bitumen coverage for the base binder was 15%. It then increased to 60% when 2% of GNPs were added. For the binder modified with 4% of GNPs, it jumped to 70%. The extremely small size of the GNPs gives them a larger surface area compared to their parent material, which results in the increase of their affinity to form a strong bond. Thus, GNPs have the ability to absorb more free asphalt binder and maximize the quantity of the structural asphalt [26]. The results prove that the addition of GNPs leads to an increased adhesion between the binder and aggregate, making the asphalt highly resistant to moisture.

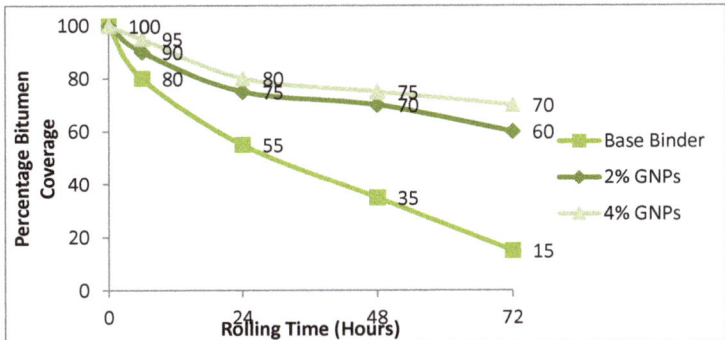

Figure 12. The Rolling Time vs the Percentage of Bitumen Coverage. GNPs = Graphene Nano-Platelets.

Figure 13. The bitumen/aggregate affinity after 72 h of rolling: (**a**) The base binder, (**b**) 2% Graphene Nano-Platelet (GNP)-modified binder and, (**c**) 4% GNP-modified binder.

3.6. Bitumen Bond Strength

The idea of the PATTI was first conceived by the paint industry. It determines the POTS of the binder while keeping a constant loading rate of 0.69 MPa/s. Figure 14 shows the testing assembly. It has been learned, from the results presented in Table 7, that an increase in additive content produces a higher Bitumen Bond Strength. When 2% of GNPs were added to the binder, the POTS increased significantly from 8.76 MPa to 11.95 MPa. With the addition of 4% of GNPs, it further increased to 13.66 MPa. The mode of failure for all of the samples was cohesive, i.e., bitumen-bitumen interface breakage. The reason behind the increment in the tensile strength of the binder can be attributed to the increased surface texture of the GNP-modified asphalt binder [29]. This is because a rough binder surface is obtained when nanoparticles are added to it, due to more particle interlocking. The elasticity of the nanoparticles also plays a part in enhancing the tensile strength of the binder. Thus, it is evident that the GNPs improve the adhesive and cohesive bond strength of the asphalt binder.

Figure 14. The Pneumatic Adhesion Tensile Test Instrument (PATTI) Assembly.

Table 7. The Pull-Off Tensile Strength (POTS) of the binder. GNPs = Graphene Nano-Platelets.

Samples	Pull-Off Tensile Strength (POTS) (MPa)
Base Binder	
1	8.67
2	8.76
3	8.73
2% GNP-Modified Binder	
1	11.67
2	11.95
3	11.67
4% GNP-Modified Binder	
1	13.66
2	13.38
3	13.38

3.7. Surface Texture of the Binder-Coated Aggregates

The images to study the surface texture of aggregates coated with GNP-modified asphalt binder were taken both from a digital camera and an SEM. Figure 15 shows the images of a piece of aggregate coated with base binder. Figure 16 shows the images of GNP-modified asphalt binder.

Figure 15. A piece of aggregate coated with base binder.

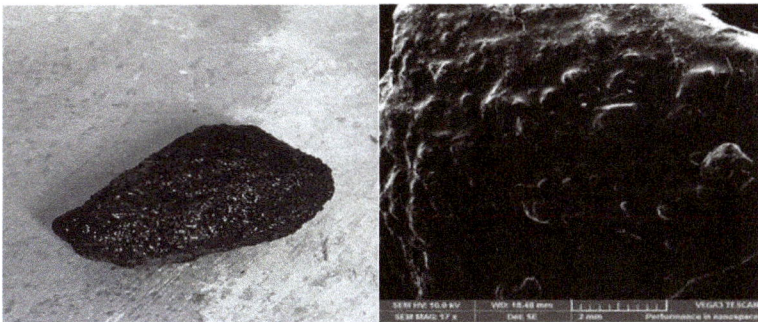

Figure 16. A piece of aggregate coated with Graphene Nano-Platelet (GNP)-modified binder.

It can be observed from the images that the surface of the aggregate coated with GNP-modified binder is rough in comparison to the one coated with the base binder. This is because nanoparticles add their nano-texture to the surface, which is also the reason that nanoparticles are used to make the pavement surface rough in order to increase its skid resistance [30].

3.8. Skid Resistance and the Polishing Effect

A British Pendulum Skid Resistance Tester was used to obtain the BPNs of the asphalt samples. Table 8 shows the results of the Skid Resistance test. It has been established from the results that the asphalt modified with GNPs has a higher skid resistance in comparison to non-modified asphalt. The reason behind this is the nano-texture provided by the GNPs.

Table 8. The British Pendulum Numbers (BPNs).

Number of Samples	Non-Modified Asphalt	2% Graphene Nano-Platelet (GNP)-Modified Asphalt	4% Graphene Nano-Platelet (GNP)-Modified Asphalt
British Pendulum Number (BPN) Before Wheel Tracking Test			
1	45	60	61
2	46	59	60
3	44	60	60
British Pendulum Number (BPN) After Wheel Tracking Test			
1	25	45	44
2	23	44	43
3	24	42	44
Percentage Reduction in Skid Resistance			
	47%	27%	28%

The reduction in the skid resistance due to the polishing of the asphalt surface (caused by the continuous movement of the wheel) is also lesser for GNP-doped asphalt. In the case of the non-modified asphalt, the reduction in the skid resistance was around 47%. On the other hand, the reduction in the skid resistance in the case of 2% and 4% GNP-modified asphalt was 27% and 28%, respectively. This is due to an increase in the stiffness of the GNP-modified binder which improves the structural performance of highways [18]. An improved structural performance ensures the intactness of the macrotexture, resulting in a high skid resistance [10].

Typically, BPN values are obtained in the field after the pavement has been subjected to traffic but, in this case, they were obtained in a laboratory to compare the skid resistance of GNP-doped asphalt with that of conventional asphalt. It is pertinent to mention here that the focus of this investigation was to determine the effect of the surface texture/roughness provided by the GNPs on the skid resistance of the asphalt.

4. Conclusions

In this study, Graphene Nano-Platelets (GNPs) were added to asphalt to modify its properties. Various tests were carried out to study rheology, moisture susceptibility, temperature susceptibility, bond strength, and skid resistance of the GNP-doped asphalt in comparison to conventional asphalt. Based on the results of the performance testing, the following conclusions have been drawn:

- In comparison to other nanomaterials, Graphene Nano-Platelets are easy to disperse. Unlike GNPs, other nanoparticles (i.e., CBNPs and CNTs) either require high shear mixing or solvent-based dispersion. The usage of smaller percentages of GNPs produces more pronounced results than with other nanoparticles [31]. GNPs save the cost of solvent and high shear mixing, making it possible for pavement engineers to use GNPs for asphalt modification on a larger scale.
- As per the results of conventional testing, GNPs have the potential to reduce the penetration value of asphalt binder by up to 48% and increase its softening point by up to 19%. The enhancement in these properties can be attributed to the small size and large surface area of GNPs, which aids in stronger bonding. The lower softening point values of the locally produced non-modified binders make them susceptible to rutting during the summers. Upon the addition of 4% of GNPs, the softening point increases to 57 °C, which is desirable due to the high temperature in Pakistan.
- According to the result of the storage stability test, GNP-modified asphalt binder is stable and can be stored for longer periods without the settling down of the nanoparticles. This makes it suitable for use in the pavement industry.
- Upon the addition of 4% of GNPs, PG 70 is achieved, which caters to the environmental conditions of Pakistan [13].
- The study of the rheology of the binder shows a significant increase in the stiffness properties of the GNP-modified binder. A maximum increase in the complex shear modulus and a decrease in the phase angle is recorded when 4% of GNPs are added to the binder. The Superpave rutting factor also increases for the GNP-doped asphalt binder, suggesting a better performance at a high temperature. The elastic nature of the GNPs contributes to this improvement.
- The optimum binder content (OBC) increases with an increase in the content of the GNPs. This trend shows that the initial cost of GNP-modified asphalt would be slightly more than conventional asphalt. But, by observing the modified asphalt performance in the results, we can conclude that doping asphalt with GNPs would reduce the life cycle cost of the pavements. This is due to the fact that pavements in Pakistan fail prematurely. A highway section designed for 20 years sometimes fails within the first two to three years of service because of excessive deformations during summers. GNP-doped asphalt pavements would require far less frequent maintenance or reconstruction cycles, thus positively affecting the life cycle costs.
- The addition of GNPs significantly reduces the temperature susceptibility and increases the resistance to permanent deformation. GNP-modified asphalt shows around a 35% reduction in the rut depth at a high temperature. At 55 °C, the wheel-tracking slope (WTS) decreased from 0.58 to 0.35 when 4% of GNPs were added to the asphalt binder. This reduction in the WTS is also suggestive of high resistance to permanent deformation.
- GNPs act as supporting material in asphalt and give it strength, similar to a steel reinforcement in concrete. This leads to an increase in the dynamic modulus of the asphalt. The dynamic modulus of GNP-modified asphalt is around 21% higher than that of non-modified asphalt, which represents a greater resistance to permanent deformation.
- GNPs significantly reduce the moisture susceptibility of asphalt. The addition of 2% and 4% of GNPs to the binder increases the percentage of bitumen coverage from 15% to 60% and 70%, respectively. This could be due to the high surface area of the GNPs, which lets them absorb the

- free or available asphalt binder, imparting structural strength to the asphalt [30]. A reduction in the moisture susceptibility indicates high durability of the asphalt.
- The bitumen bond strength test carried out using a PATTI shows that GNPs contribute to improving the adhesive and cohesive bonding of asphalt. The addition of GNPs to binder leads to an increase in the Pull-Off Tensile Strength of around 60%. This can be attributed to the hydrogen bonds and Van der Waals forces in nano-hybrid material [31].The addition of GNPs improves an important safety parameter, the skid resistance, and also increases the asphalt resistance against polishing. The inclusion of the GNPs decreased the percentage reduction of the skid resistance due to polishing from 47% to 27%. GNPs impart nanotexture to the asphalt, which contributes to enhancing the skid resistance.

The current research on the use of nanomaterials in asphalt mixtures is being carried out in two phases. The first phase involves the laboratory characterization of selected materials. The second phase includes the field investigation through the formation of test tracks in the field exposed to the actual environmental or traffic conditions. This paper presents the findings of the first phase only. The materials shortlisted during the first phase will be subjected to field investigation in the second phase. The handling and transportation of a stiff binder in the field will also be studied. Enhancing the skid resistance of asphalt using GNPs is a new concept and needs further exploration. This will also be carried out in the second phase.

Author Contributions: Conceptualization, N.A.; Data curation, M.H., J.R., and M.F.u.H.; Formal analysis, M.H., N.A., M.A.K., J.R., M.F.u.H., and J.; Investigation, M.H.; Methodology, M.H. and M.A.K.; Supervision, N.A. and M.A.K.; Writing—original draft, M.H.; Writing—review and editing, N.A., M.A.K., J., S.B.A.Z., and M.A.N. All of the authors approved and studied the final paper.

Funding: This research received no external funding.

Acknowledgments: The authors would like to acknowledge the support of the Civil Engineering department at the University of Engineering and Technology, Taxila.

Conflicts of Interest: The authors declare no conflict of interest.

References

1. Chong, D.; Wang, Y.; Zhao, K.; Wang, D.; Oeser, M. Asphalt Fume Exposures by Pavement Construction Workers: Current Status and Project Cases. *J. Constr. Eng. Manag.* **2018**, *144*, 05018002. [CrossRef]
2. Shen, X.; Wang, Z.; Wu, Y.; Liu, X.; He, Y.; Kim, J. Multilayer Graphene Enables Higher Efficiency in Improving Thermal Conductivities of Graphene/Epoxy Composites. *Nano Lett.* **2016**, *16*, 3585–3593. [CrossRef]
3. Sharafimasooleh, M.; Shadlou, S.; Taheri, F. Effect of functionalization of graphene nanoplatelets on the mechanical response of graphene/epoxy composites. *J. Mater. Des.* **2014**, *66*, 142–149. [CrossRef]
4. Wang, F.; Drzal, L.T. Development of Stiff, Tough and Conductive Composites by the Addition of Graphene Nanoplatelets to Polyethersulfone/Epoxy Composites. *Materials* **2018**, *11*, 2137. [CrossRef]
5. Meng, Q.; Kuan, H.; Araby, S.; Kawashima, N. Effect of Interface Modification on PMMA/Graphene Nanocomposites. *J. Mater. Sci.* **2014**, *49*, 5838–5849. [CrossRef]
6. Han, M.; Li, J.; Muhammad, Y.; Hou, D.; Zhang, F.; Yin, Y. Effect of polystyrene grafted graphene nanoplatelets on the physical and chemical properties of asphalt binder. *Constr. Build. Mater.* **2018**, *174*, 108–119. [CrossRef]
7. Li, J.; Han, M.; Muhammad, Y.; Liu, Y.; Yang, S.; Duan, S. Comparative analysis, road performance and mechanism of modification of polystyrene graphene nanoplatelets (PS-GNPs) and octadecyl amine graphene nanoplatelets (ODA-GNPs) modified SBS incorporated asphalt binders. *Constr. Build. Mater.* **2018**, *193*, 501–517. [CrossRef]
8. Le, J.; Marasteanu, M.; Turos, M. *Graphene Nanoplatelet (GNP) Reinforced Asphalt Mixtures: A Novel Multifunctional Pavement Material*; NCHRP IDEA Project; Transportation Research Board: Washington, DC, USA, 2016.

9. Brcic, H. Investigation of the Rheological Properties of Asphalt Binder Containing Graphene Nanoplatelets, Norwegian University of Science and Technology, Department of Civil and Transport Engineering. Master's Thesis, Norwegian University of Science and Technology, Trondheim, Norway, 2016.
10. Vaiana, R.; Capiluppi, G.F.; Gallelli, V.; Iuele, T.; Minani, V. Pavement Surface Performances Evolution: An Experimental Application. *Procedia Soc. Behav. Sci.* **2012**, *53*, 1149–1160. [CrossRef]
11. Ferrari, A.C. Raman Spectroscopy of Graphene and graphite: Disorder, Electron-phonon coupling, Doping and Nonadiabatic Effects. *Solid State Commun.* **2007**, *143*, 47–57. [CrossRef]
12. Hussan, S.; Kamal, M.A.; Hafeez, I.; Ahmad, N.; Khanzada, S.; Ahmed, S. Modelling Asphalt Pavement Analyzer Rut Depth Using Different Statistical Techniques. *Road Mater. Pavement Des.* **2018**, 1–26. [CrossRef]
13. Mirza, M.W.; Abbas, Z.; Rizvi, M.A. Temperature Zoning of Pakistan for Asphalt Mix Design. *Pak. J. Eng. Appl. Sci.* **2011**, *8*, 49–60.
14. Khan, K.M. Development of Superpave Performance Grading Map for Pakistan. *Life Sci. J.* **2013**, *10*, 355–362.
15. NHA General Specification. *Prepared by SAMPAK International (Pvt.) Ltd.*; National Highway Authority: Lahore, Pakistan, 1998.
16. Zou, X.; Sha, A.; Ding, B.; Tan, Y.; Huang, X. Evaluation and Analysis of Variance of Storage Stability of Asphalt Binder Modified by Nanotitanium Dioxide. *Adv. Mater. Sci. Eng.* **2017**, *2017*, 6319697. [CrossRef]
17. Abdullah, M.E.; Tun, U.; Onn, H.; Zamhari, K.; Nayan, N.; Tun, U.; Onn, H.; Hainin, M.R. Storage stability and physical properties of asphalt modified with nanoclay and warm asphalt additives. In Proceedings of the Nineteenth Annual International Conference on COMPOSITES/NANO ENGINEERING (ICCE-19), Shanghai, China, 24–30 July 2011.
18. Muniandy, R.; Yunus, R.; Salihudin, H.; Aburkaba, E.E. Effect of Organic Montmorillonite Nanoclay Concentration on the Physical and Rheological Properties of Asphalt Binder. *Aust. J. Basic Appl. Sci.* **2013**, *7*, 429–437.
19. Sarsam, S.I. Effect of Nano Materials on Asphalt Cement Properties. *Int. J. Sci. Res. Knowl.* **2013**, *1*, 422. [CrossRef]
20. Mashaan, N.S.; Ali, A.H.; Karim, M.R.; Abdelaziz, M. Effect of crumb rubber concentration on the physical and rheological properties of rubberised bitumen binders. *Int. J. Phys. Sci.* **2011**, *6*, 684–690.
21. Nazari, H.; Naderi, K.; Moghadas, F. Improving aging resistance and fatigue performance of asphalt binders using inorganic nanoparticles. *Constr. Build. Mater.* **2018**, *170*, 591–602. [CrossRef]
22. Sadeghnejad, M.; Gholamali, S. Experimental Study on the Physical and Rheological Properties of Bitumen Modified with Different Nano Materials (Nano SiO_2 & Nano TiO_2). *Int. J. Nanosci. Nanotechnol.* **2017**, *13*, 253–263.
23. Zhang, H.; Gao, Y.; Guo, G.; Zhao, B.; Yu, J. Effects of ZnO particle size on properties of asphalt and asphalt mixture. *Constr. Build. Mater.* **2018**, *159*, 578–586. [CrossRef]
24. Mrugała, J.; Chomicz-kowalska, A. Influence of the production process on the selected properties of asphalt concrete. *Procedia Eng.* **2017**, *172*, 754–759. [CrossRef]
25. Faizan, M.; Ahmad, N.; Nasir, M.A.; Hafeez, M.; Rafi, J.; Bilal, S.; Zaidi, A.; Haroon, W. Carbon Nanotubes (CNTs) in Asphalt Binder: Homogeneous Dispersion and Performance Enhancement. *Appl. Sci.* **2018**, *8*, 2651. [CrossRef]
26. Cai, L.; Shi, X.; Xue, J. Laboratory evaluation of composed modified asphalt binder and mixture containing nano-silica/rock asphalt/SBS. *Constr. Build. Mater.* **2018**, *172*, 204–211. [CrossRef]
27. Li, P.; Zheng, M.; Wang, F.; Che, F.; Li, H.; Ma, Q.; Wang, Y. Laboratory Performance Evaluation of High Modulus Asphalt Concrete Modified with Different Additives. *Adv. Mater. Sci. Eng.* **2017**, *2017*, 7236153. [CrossRef]
28. Yao, H.; You, Z. Effectiveness of Micro-and Nanomaterials in Asphalt Mixtures through Dynamic Modulus and Rutting Tests. *J. Nanomater.* **2016**, *2016*, 2645250. [CrossRef]
29. Leiva-Villacorta, F.; Aguiar-Moya, J.P.; Salazar-Delgado, J.; Loría-Salazar, L.G. Adhesion Performance of Nano-Silica Modified Binder. In Proceedings of the 95th Annual Meeting of the Transportation Research Board, Washington, DC, USA, 10–14 January 2016.

30. Hsieh, C.; Chen, J.; Kuo, R.; Lin, T.; Wu, C. Influence of surface roughness on water- and oil-repellent surfaces coated with nanoparticles. *Appl. Surf. Sci.* **2005**, *240*, 318–326. [CrossRef]
31. Rafi, J.; Kamal, M.A.; Ahmad, N.; Hafeez, M.; Faizan, M.; Asif, S.A.; Shabbir, F.; Bilal, S.; Zaidi, A. Performance Evaluation of Carbon Black Nano-Particle Reinforced Asphalt Mixture. *Appl. Sci.* **2018**, *8*, 1114. [CrossRef]

© 2019 by the authors. Licensee MDPI, Basel, Switzerland. This article is an open access article distributed under the terms and conditions of the Creative Commons Attribution (CC BY) license (http://creativecommons.org/licenses/by/4.0/).

Article

Effect of Chemical Composition of Bio- and Petroleum-Based Modifiers on Asphalt Binder Rheology

Punit Singhvi [1,*], Javier J. García Mainieri [1], Hasan Ozer [2], Brajendra K. Sharma [3] and Imad L. Al-Qadi [1]

- [1] Illinois Center for Transportation, University of Illinois at Urbana-Champaign, Champaign, IL 61820, USA; jjg5@illinois.edu (J.J.G.M.); alqadi@illinois.edu (I.L.A.-Q.)
- [2] School of Sustainable Engineering and the Built Environment, Arizona State University, Tempe, AZ 85281, USA; hasan.ozer@asu.edu
- [3] Prairie Research Institute–Illinois Sustainable Technology Center, University of Illinois Urbana–Champaign, Champaign, IL 61820, USA; bksharma@illinois.edu
- * Correspondence: singhvi3@illinois.edu

Received: 4 March 2020; Accepted: 3 May 2020; Published: 7 May 2020

Featured Application: This study provides guidelines for asphalt binder modifier selection to produce low modulus binders (softer binders) to meet desired quality. The research focuses on reducing expected cracking susceptibility of modified asphalt binders after long-term aging. The study recommends the development of engineered modifiers for specific paving applications.

Abstract: In recent years, increased use of recycled asphalt materials (RAP) has created a need for softer binders to compensate stiffer binder coming from RAP. Economic alternatives, like recycled oils and proprietary bio-based oils, can be potential modifiers that will reduce the dependence on petroleum-based alternatives. However, there is limited information on the long-term rheological performance of binders modified with proprietary modifiers. These modifiers are chemically complex and their interaction with binders further complicates the binder chemistry. Therefore, the objective of this study was to evaluate the impact of modifier chemistry on modified binders' long-term cracking potential. A base binder of Superpave Performance Grade (PG) 64-22 was used to develop PG 58-28 binder using six different modifiers. An unmodified PG 58-28 was included for a comparative analysis. A few modified binders rheologically outperformed the base binder and others performed similarly. The modifier derived from recycled engine oil showed the worst performance. Chemical analysis indicated that the best performing modified binders had significant amounts of nitrogen in the form of amines. On the other hand, poor performing modified binder had traces of sulfur. Additionally, modifiers with lower average molecular weights appeared to have a positive impact on the performance of aged binders.

Keywords: Asphalt modification; modifier chemistry; long-term aging; asphalt rheology; phase angle; delta T_c

1. Introduction

Asphalt concrete (AC) is one of the most commonly used pavement materials in the United States and worldwide. More than 400 million tons of AC are produced in the US annually, which requires 20 million tons of asphalt binder [1]. AC mixture is a heterogeneous composite of asphalt binder, mineral aggregates, and air voids. The performance of AC mixtures is greatly affected by aggregate characteristics, binder chemistry and rheology, mixture volumetrics, and aging. Aging of AC is a continuous process; it is dominated by volatilization and oxidation of asphalt binder in the short- and long-term, respectively. Volatilization in binders refers to loss of lighter fractions when exposed to high temperatures and occurs during the production and construction stages. Oxidation in binders is caused by photo-oxidation and thermal oxidation during pavement's service-life [2]. Aging increases AC brittleness that may result in cracking. Hence, binders with superior aging resistance characteristics may delay AC cracking and increase pavement service life.

Asphalt binder (AB) is an important component for the construction of AC pavements. It is currently produced from the fractional distillation of crude petroleum at refineries. In recent years, the increasing use of harder and aged recycled asphalt materials in pavement applications has significantly increased the need for softer ABs [1]. Logistical limitations and the cost of refineries to produce softer ABs instead of products of higher financial value, result in a shortage of soft straight-run or unmodified AB in the market [3].

To overcome the current demand for softer ABs, traditional petroleum-based "softeners", like AB flux and aromatic oils are blended with straight-run AB [4]. Blending "softeners" with readily available products, like recycled oils and bio-based oils, provides an opportunity to manufacture required AB economically and reduce the dependence on petroleum-based products.

A variety of proprietary products are available to modify AB to achieve softer grades. They are used with limited knowledge about their long-term performance. Poor durability and/or extended cracking issues have been identified when re-refined engine oil bottoms (ReOB) and waste engine oils (WEO) have been blended in AB [5–9]. On the other hand, use of certain bio-based oils in ABs increases the oxidation potential which makes pavements vulnerable to cracking over the long-term [10]. This limits or restricts the use of these products. However, there are some specialized bio-based softeners that have shown enhanced long-term performance of ABs and reduced cracking potential [11]. Therefore, the potential of bio-based modifiers to produce softer PGs is investigated in this study. A ReOB-based modifier was included in this study for comparison purposes.

Superpave Performance Grading (PG) is a rheology-based system, currently used to trade and specify AB in the United States. The PG system has limitations in identifying long-term performance of modified asphalt binder (MAB). The use of bio-based modifiers can significantly impact AB chemistry and rheology without changing its Superpave grading. The current challenges for the market-entry of MAB are: (i) their complex chemistry, (ii) their uncertain long-term rheological performance, and (iii) the lack of a robust grading system that can discriminate them to ensure long-term performance.

Low-temperature performance of AB is critical to prevent adverse effects of thermal cracking. Use of recycled asphalt pavements (RAP)—obtained from milling of old asphalt layers—and recycled asphalt shingles (RAS)—typically obtained from either tear-off shingles or manufactured waste shingles—has increased in AC pavement and requires adequately-performing binders to avoid premature cracking [12–18]. The ΔT_c parameter has been used to assess the low-temperature induced cracking performance of AB [19,20]. Low-temperature ductility of AB has been related to pavement cracking performance [21]. Glover et al. [22] developed a rheological parameter based on dynamic shear rheometer (DSR) frequency sweep tests, that strongly correlates to low-temperature ductility; a simplification of this parameter is known as the Glover–Rowe (GR) parameter [23]. Literature provides ΔT_c and GR parameters as good low-temperature cracking indices [19]. Therefore, frequency sweeps and bending beam rheometer (BBR) tests were considered in this study to obtain them and evaluate asphalt binder's expected cracking performance.

Limitations of using Superpave's $|G^*|\sin\delta$ to assess intermediate-temperature cracking susceptibility of asphalt binders are well reported [3,24]. Black space diagrams may provide insights to the rheological properties that drive cracking susceptibility at intermediate temperature.

Asphalt cracking is further aggravated with aging. Field aging depends on geographical location and environmental factors and varies along the pavement depth. The current pressure aging vessel (PAV) aging for 20 h at 90, 100, or 110 °C and a pressure of 2.1 MPa is not sufficient to represent realistic long-term aging of binders [22,25]. Meanwhile, researchers are investigating 40-h PAV as an alternate solution [26]. Aging up to 60-h PAV has also been used to evaluate AB long-term performance [27]. Laboratory aging conditions of 20-h, 40-h, and 60-h PAV were investigated in this study.

Chemical composition of AB plays an important role in durability of asphalt pavements [28]. Modifiers' chemical composition can affect AB compatibility and susceptibility to oxidation, changing its rheological properties and long-term performance [3,27,29,30]. Chemical characterization of ABs and MABs was conducted, in many studies, using elemental analysis, gel permeation chromatography (GPC), Fourier-transform infrared spectroscopy (FTIR), and thin-layer chromatography flame ionization detection (TLC-FID). Carbonyl and sulfoxide indices from FTIR have been widely used to track oxidation and evaluate the impacts of long-term aging on binder characteristics [31–35]. Molecular weight tends to increase with aging in binders and has been reported by several authors [30,33,36–38]. Asphalt binders are composed of maltenes and asphaltenes and their interaction with aging drives the mechanical properties of the binder [28,30,38–41].

There is a need for understanding the impact of various modifiers on ABs and their resulting long-term field performance. Simple laboratory protocols with reasonable and consistent predictive capabilities of field performance are essential to optimize modifier selection and dosage. Therefore, the objective of this study was to evaluate modifiers' chemical characteristics and their impact on MAB's long-term cracking potential. The fundamental relationship between chemical composition of modifiers and its effect on MAB's rheological properties was investigated.

2. Materials and Methods

2.1. Materials

Various sources of bio-based products were selected to produce "softer" MABs with equivalent or better long-term performance as the binder obtained from crude oil sources. Softer binders are defined as binders with lower modulus and lower PG. Such binders are commonly used to control cracking in colder climates or neutralize RAP's relatively stiff binder. PG 58-28 was the target binder grade to be produced using commonly available PG 64-22 as base binder, referred to as S1. The base binder was selected and sampled from a refinery terminal in Illinois, USA. Five bio-based AB modifiers and one ReOB modifier available in the US industry were procured for modification. The names of the modifiers used in this study are kept confidential and are designated as shown in Table 1 from hereon. Table 1 lists the modifier type as provided by the suppliers. MABs were labelled with base AB's designation and modifier's designation followed by dosage (% by weight) of modifier in the blend, e.g., S1-A-3.5 represents MAB obtained from the base binder S1 blended with 3.5% (% by weight) of modifier "A". A softer unmodified binder, PG 58-28 (labelled as S5), was also included in the study as a benchmark for modified binders.

Table 1. Modifiers used in this study and their classification.

Modifier	Type [1]
A	NA [2]
C	Fatty acid derivative
D	Bio-oil blend
E	Modified vegetable oil
G	Glycol amine
K	ReOB

[1] as reported by suppliers (from US); [2] not provided by the suppliers.

2.2. Asphalt Binder Modification

Modifier blending methodology, binder heating cycles, temperatures applied during splitting process, and storage are discussed in this section. A high shear mixer (Cafarmo BDC1850) with a Heidolph PR31 ringed propeller (33-mm diameter fan) was used for blending. The blending was performed at a steady temperature of 130 ± 10 °C. The temperatures were maintained using a Glas-Col

LLC heating mantle capable of handling 1-L capacity aluminum can. The methodology is illustrated in Figure 1.

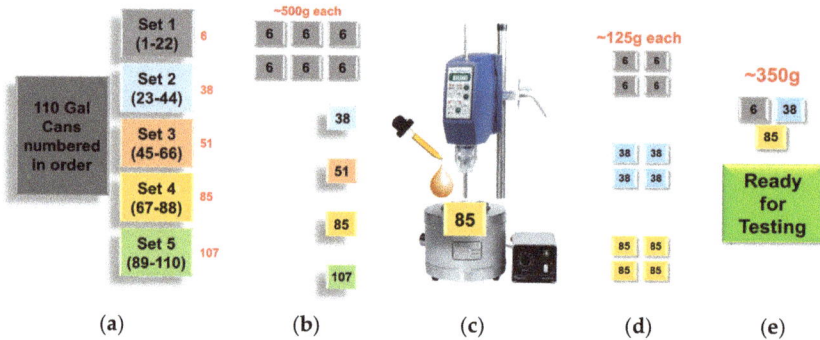

Figure 1. Modifier blending approach (**a**) labelled cans of base binder divided in five sets; (**b**) 3.8 L cans split into 6–1.0 L cans; (**c**) shear mixer for blending and modifier addition; (**d**) modified binder split into 4–240 mL cans; and (**e**) 3–240 mL cans from different sets combined to produce materials for all aging conditions.

The following were the steps for modifying binders used for chemical and rheological characterization:

1. PG 64-22 binder was sampled from the refinery terminal in 110–3.8 L cans. They were labelled from 1 to 110 in the order they were collected. The base binder was then grouped in five different sets as shown in Figure 1a.
2. Binder from each set was split to 6 cans of around 500 g each (Figure 1b). Each can was heated to 135 °C for 2 h to reach a flowing state and was stirred prior to splitting. The split samples were then stored for modification.
3. Before modification, the split base binder (S1) was heated for 30 min in a heating mantle (Figure 1c) to reach 130 ± 10 °C to ensure that material was steadily flowing prior to blending.
4. The modifier was added (weight measured with a 0.01 g readability scale), the propeller was inserted half-way into the depth of the material and was set to 1000 rpm for 20 min. During blending, the temperature was maintained at 130 ± 10 °C and formation of vortices was prevented to avoid air bubbles in the blend.
5. After blending, the MAB was divided into 4–240 mL cans (Figure 1d).
6. Steps 3 to 5 were repeated for the binders from different sets mentioned in Step 1 for obtaining representative samples.
7. Later, 3–240 mL cans from different batches of each MAB were combined to proceed with aging conditioning (Figure 1e).

Aging conditions used in the study were: Unaged (UA), Rolling Thin-Film Oven (RTFO) in accordance with American Association of State Highway and Transportation Officials (AASHTO) specification T240-13, 20-h Pressure Aging Vessel (PAV) including vacuum degassing and in accordance with AASHTO specification T28-12, 2PAV and 3PAV. 2PAV and 3PAV conditions were obtained by running continuous 40-h and 60-h PAV cycles, respectively. Once samples reached their required aging conditions, they were stored in small 30 mL cans until tested to avoid multiple heating cycles. To avoid changes in chemical and rheological properties, cans once heated for testing were not re-used. Same treatment was given to the unmodified base binder S1 and unmodified reference binder S5.

2.3. Modifier Chemistry Experimental Program

2.3.1. Elemental Analysis

Elemental analyses of modifiers were conducted in an Exeter Analytical (Chelmsford, MA, USA) CE-440 elemental analyzer. The proportions of carbon, hydrogen, nitrogen, and sulfur (CHNS) elements in the material's composition were expressed in percent. The proportion of oxygen (O) was obtained by subtracting CHNS percentage from 100.

2.3.2. Chemical Functional Groups

Thermo Nicolet Nexus 670 FTIR spectrometer was used to detect the chemical functional groups present in modifiers in the range of wavenumbers 600–4000 cm^{-1}. Data was collected at a resolution of 4 cm^{-1} with number of scans set to 128. Three replicates were tested for each modifier. The method was based on attenuated total reflection (ATR).

2.3.3. Molecular Weight Analysis

The molecular weight analysis was conducted using GPC. The system consists of a Waters 2695 separation module connected to two Styragel HR1 SEC columns (7.8 mm × 300 mm) in series followed by a Waters 2414 RI detector and a computer with Empower Pro and data acquisition software. Samples of 3% w/w were prepared in tetrahydrofuran (THF), a carrier solvent with a flow rate of 1.0 mL/min and an injection volume of 20 µL; they were filtered using a 0.45 µm millipore polytetrafluoroethylene (PTFE) syringe filter to remove suspended particulates. To detect analytes, a constant flow of fresh eluent was supplied to the column via a pump.

The resulting chromatographic data was processed for number-average molecular weight (M_n), weight-average molecular weight (M_w), and polydispersity index (PDI) using Equations (1)–(3), respectively. The molecular weights were calculated based on the component molecular weights (M_i) determined from the retention time calibration curve and signal intensities (N_i).

$$M_n = \frac{\sum M_i N_i}{\sum N_i} \tag{1}$$

$$M_w = \frac{\sum M_i^2 N_i}{\sum M_i N_i} \tag{2}$$

$$PDI = \frac{M_w}{M_n} \tag{3}$$

The retention time calibration curve was developed by fitting log-scale molecular weights to their retention time for standard material with known molecular weights using a 3-degree polynomial. The fitted curve was then used to measure the molecular weights of unknown modifiers using the chromatographic data. The distributions with shorter retention times correspond to larger molecular size whereas longer retention times represent smaller sizes. The molecular weights are reported in Daltons.

2.3.4. Binder Fractionation

Two percent (weight by volume) solutions of the modifiers were prepared in dichloromethane and filtered through a 0.45 µm millipore PTFE syringe filter to remove insoluble suspended particles from the solution. The suspended particles are referred to as residue from hereon. The sample solution (1 µL) was spotted on chromrods coated with a thin film of silica gel, using a microsyringe. The separation of bitumen into four generic fractions: saturates, aromatics, resins, and asphaltenes (SARA) was performed in a three-stage development process using n-heptane, toluene, and THF. The chromrods were dried for 10 min and humidified in NaNO$_2$ for 10 min between each development. The chromrods were scanned with an Iatroscan MK-5 analyzer (Iatron Laboratories Inc., Tokyo, Japan) with a flame ionization detection (FID), which provided chromatograms with peaks for SARA composition.

One of the used modifiers (G) was insoluble in dichloromethane. However, the modifier dissolved in water, acetone and methanol and hence the sample was prepared in methanol for performing the test.

2.4. Binder Rheology Experimental Program

Rheological characterization of binders included determining Superpave performance grade, ΔT_c, and frequency sweep test parameters using a dynamic shear rheometer (DSR). Two replicates were tested for each rheological test and an average value was reported. AASHTO-intra-laboratory precision limits were obeyed. Specifically, in the case of frequency sweeps, the coefficient of variation was limited to 7% on any complex modulus measurement. All DSR measurements were performed in a Kinexus KNX2712 equipment with an active hood for temperature control while the BBR measurements were performed on a Cannon instrument.

The modifier dosage was selected to achieve similar PG for all modified binders in this study. Superpave system of grading is the current method of selecting binders in the US. Even though the binders tested for this study (except the base binder and one of the modified binders) would be assigned the same grading by this system, their chemistry could affect their rheological characteristics and susceptibility to oxidation; hence, it may affect their long-term performance. Given that the same base binder (S1) was used, the differences reported in the study's experimental program were caused by the modifiers.

2.4.1. Superpave Performance Grading (PG)

All the binders used in the study were tested for Superpave PG. The tests were performed in accordance with AASHTO specifications: T315-19 (DSR) and T313-19 (BBR). Continuous PGs (true grades) were also determined.

2.4.2. ΔT_c Parameter

BBR measurements for PAV-aged samples were obtained for PG. Additionally, BBR measurements were recorded for all samples at 2PAV-aged and 3PAV-aged conditions in accordance with AASHTO T313-19 specification.

ΔT_c was computed using the following Equation (4):

$$\Delta T_c = PG_{Stiffness} - PG_{m-value} \qquad (4)$$

where $PG_{Stiffness}$ and $PG_{m-value}$ are the temperatures at which samples pass PG criteria for stiffness and slope of stiffness curve (relaxation), respectively. $PG_{m-value}$ at all aging conditions and $PG_{Stiffness}$ value at PAV condition (except for K-modified binder) were interpolated as stated in [20]. $PG_{Stiffness}$ value at 2PAV and 3PAV conditions, and K-modified binder at all aging conditions, were extrapolated using the same equation.

2.4.3. Frequency Sweep Test

Frequency sweeps were performed after short conditioning (10 cycles of 0.1% strain at 15 °C and a frequency of 0.5–0.61 rad/s). Complex shear modulus (G*) and phase angle (δ) were obtained at all frequencies for all samples at UA, RTFO, PAV, 2PAV, and 3PAV. Data was measured in isotherms of 15 °C, 25 °C and 35 °C. An additional 5 °C isotherm was included for UA and RFTO-aged samples to ensure crossover (when phase angle δ is 45 degrees) data were measured. Eighteen data points were collected per isotherm in the frequency range from 1.00 rad/s to 62.83 rad/s at constant shear strain of 1.6% for UA, 1.2% for RTFO and PAV-aged samples, and 1.0% for 2PAV and 3PAV samples.

These strains were selected to ensure measurements within linear viscoelastic range (LVER) of the samples. Harmonic distortion between strain excitations and stress responses was lower than 1% for all measurements. In addition, measured torque was in the operational range of equipment for all sweeps performed. Additionally, a built-in sequence to verify whether measurements for each sample were taken in the LVER was incorporated before the isotherms.

Data from different isotherms were then manually shifted to match G* at 15 °C (reference temperature) to create master curves. Polynomial fitting was performed to obtain the presented black space diagrams.

Glover–Rowe Parameter (GR)

GR parameter was computed using Equation (5):

$$GR = \frac{G^*(\cos\delta)^2}{\sin\delta} \tag{5}$$

G^* and δ are the complex shear modulus and phase angle at 15 °C and 0.005 rad/s. G^* and δ at 15 °C and 0.005 rad/s were obtained from measured data. In some cases, the data were extrapolated using polynomial-fits from master curves and black space diagrams.

3. Results

This section presents the results from chemical compositional testing of the modifiers and rheological testing of the modified and base binders.

3.1. Chemical Characterization of Modifiers

3.1.1. Elemental Analysis

The results from elemental analysis are presented in Table 2. Modifiers A, D, and E show similar elemental composition of carbon, hydrogen, nitrogen and oxygen with additional sulfur in E (0.33%). Modifier K has slightly higher carbon (79.7%) and hydrogen (12.7%) content but is lower in oxygen (5.7%) compared to A, D, and E. Modifier K also has higher sulfur (0.98%) content than E. Modifier C has relatively higher nitrogen (3.5%) and oxygen (14.9%) and is low on Carbon (70.1%) in comparison to A, D, E, and K. Modifier G possesses very different composition compared to all other modifiers. Its elemental oxygen (33.1%) and nitrogen (9.0%) contents are the highest while the carbon (47.9%) is the lowest compared to other modifiers.

Table 2. Elemental analysis of modifiers.

Sample	C (%)	H (%)	N (%)	O (%)	S (%)
A	77.3	11.9	0.4	10.4	0.001
C	70.1	11.5	3.5	14.9	0.004
D	76.1	11.8	0.8	11.3	0.03
E	77.1	11.5	0.6	9.9	0.33
G	47.9	10.0	9.0	33.1	0.00
K	79.7	12.7	0.9	5.7	0.98

3.1.2. Fourier Transform Infrared Spectroscopy (FTIR)

Figure 2a shows the full FTIR spectra for all modifiers, from which chemical functional groups present in the modifiers can be identified. The majority of the absorbance peaks were observed in wavenumbers ranging from 1000 to 1800 cm^{-1} (Figure 2b) and 2700 to 3200 cm^{-1} (Figure 2c).

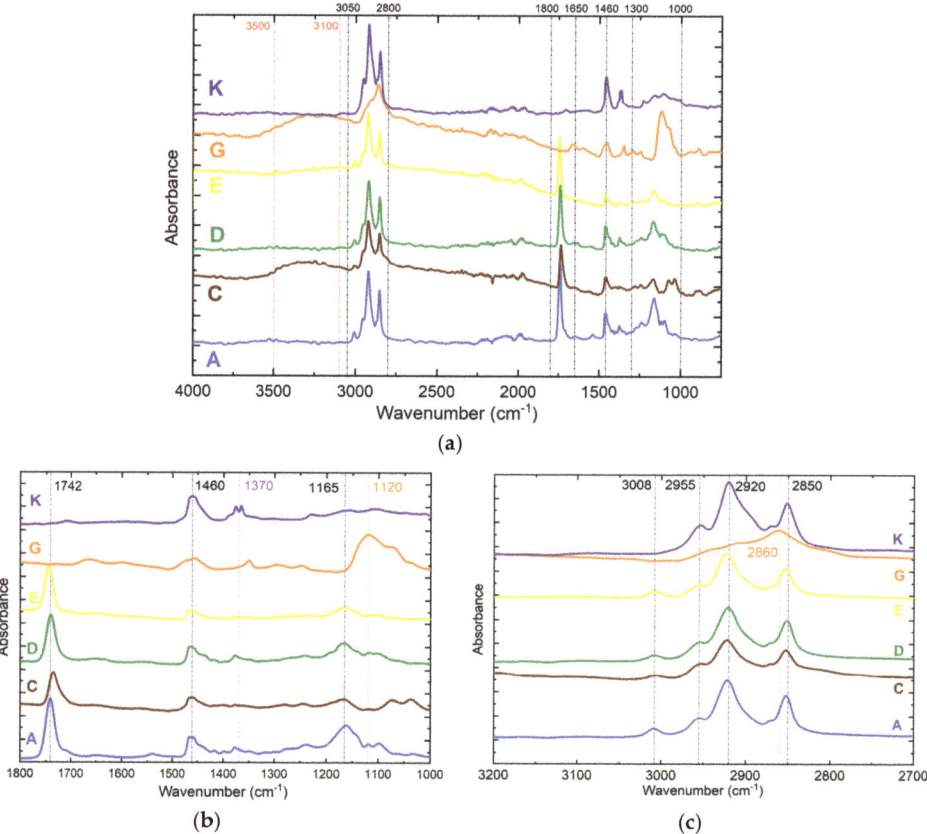

Figure 2. FTIR spectra for modifiers from (**a**) 600–4000 cm^{-1}; (**b**) 1000–1800 cm^{-1}; and (**c**) 2700–3200 cm^{-1}.

Following are the observations from the FTIR spectra shown in Figure 2:

- All modifiers show characteristic peaks in the range of 2800–3000 cm^{-1} and 1300–1460 cm^{-1} which correspond to C-H stretching and C-H bending, respectively (Figure 2a).
- Peaks at 2850, 2920, 2955, and 3008 cm^{-1} correspond to C-H stretching and are observed in modifiers A, C, D, E, and K (Figure 2c). These peaks are absent in G. Instead, a distinct broad peak at 2860 cm^{-1} in G was observed (Figure 2c). This peak is representative of N-H stretching specific to amine salt.
- The common peak at 1460 cm^{-1} in all the modifiers shows the presence of C-H bending for alkanes (Figure 2b). It is the most common functional group present in hydrocarbons.
- Series of peaks were observed from 1000–1300 cm^{-1} in all modifiers (Figure 2a). This may reflect the presence of either alkoxy (-C-O-), phenyl (=C-O-), or C-N stretching.
- Peaks for modifiers A, C, D, and E at 1165 cm^{-1} show the presence of a common functional group which can be either of alkoxy (-C-O-), phenyl (=C-O-), or C-N (Figure 2b). Because a negligible amount of nitrogen was observed in these modifiers, the probability of alkoxy (-C-O-) or phenyl (=C-O-) groups presence is higher.
- An accentuated peak at 1120 cm^{-1} for G and a smaller peak around 1050 cm^{-1} for C might correspond to C-N stretching (Figure 2b). This is characteristic of amine stretching (1000–1250 cm^{-1}).

- Smaller peaks at 1350 and 1375 cm^{-1} for modifier K correspond to stretching of sulfoxide (-S=O) group (Figure 2b).
- Strong peaks of carbonyl (-C=O) stretching were observed in modifiers A, C, D, and E between 1650–1800 cm^{-1}. The peaks are close to wavenumber 1742 cm^{-1} which may reflect the presence of aldehydes, ketones, esters, or carboxylic groups (-C=O). However, a very small peak for modifier G was also observed closer to wavenumber 1650 cm^{-1} which is characteristic of amides (-NC=O) (Figure 2b). There is no peak for modifier K in this region.
- Modifier G and C have a single broad peak in range of 3100–3500 cm^{-1} (Figure 2a). This peak is usually due to the stretching in alcohols (O-H) or secondary amines (-R$_1$R$_2$N-H). Based on elemental analysis, G and C have higher nitrogen content compared to other modifiers. Therefore, presence of secondary amines is highly likely in these modifiers.
- The spectral analysis results of modifiers confirm the type as provided by the suppliers (Table 1). No information on modifier A's composition was provided; however, the FTIR results suggest that modifier A has characteristics similar to the bio-oil blend (modifier D).

3.1.3. Gel Permeation Chromatography

The average molecular weights (M_n and M_w) for modifiers A, C, D, and E ranged from 3700 to 4700, and for G it is around 2500 with polydispersity index (PDI) in a range of 1.03–1.13 as shown in Table 3. Interestingly, modifier K has a high M_n of 8933 and significantly higher M_w of 48,784 which is the reason for high PDI of 5.46. The plot showing retention times (Figure 3) suggests that even though modifiers A, C, D, and E have similar range of molecular weights, M_n and M_w (Table 3), their molecular size distribution varies. Following are the observations from GPC analysis as per Figure 3:

- The molecular weight distribution in modifier A and D showed two peaks at similar retention times but with different intensities. The larger peak occurred at lower retention times than the smaller peak, which indicates presence of two different sized molecules with higher proportion of larger size particles.
- Modifier C has a similar two peak distribution to A and D. However, the two peaks occur at longer retention times, indicating that the respective weights are smaller than A and D.
- Modifiers E and G have one sharp peak indicating the presence of single weight molecule. However, the peaks occur at different retention times which represent different particles.
- Modifier G has the largest retention time suggesting lower average molecular weight.
- Modifier K has the largest molecular weights among the modifiers. The molecular size distribution is wider compared to all other modifiers with two peaks. The wider distribution indicates a wide range of molecular size present in modifier K, resulting in a high PDI.

Table 3. Number-average molecular weight, weight-average molecular weight, and polydispersity index of modifiers.

Modifier	Number-Average Molecular Weight (M_n)	Weight Average Molecular Weight (M_w)	Polydispersity Index (PDI)
A	4450	4616	1.04
C	3709	3971	1.07
D	4345	4909	1.13
E	4684	4805	1.03
G	2349	2596	1.11
K	8933	48,784	5.46

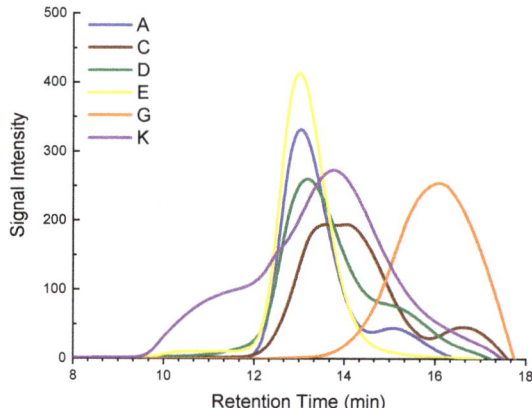

Figure 3. Signal intensity vs. retention time for molecular size determination of modifiers.

3.1.4. Thin Layer Chromatography Flame Ionization Detection (TLC-FID)

Table 4 shows the SARA composition and the unfiltered residue for different modifiers. Following are the observations from SARA analysis:

- Modifiers A, C, D, and E are primarily composed of resins.
- Modifiers A and D consist of some aromatics (A = 19.7% and D = 10.5%), limited asphaltenes (A = 4.6% and D = 7.5%), and some saturates (A = 8.3% and D = 1.1%).
- Modifiers C and E show limited or no presence of aromatics and saturates. They are primarily composed of resins (C = 70.1% and E = 83.4%) and asphaltenes (C = 23.2% and E = 13%).
- Modifier K has a significant proportion of saturates (59.3%) and exceptionally high residue content (26%) compared to other modifiers.

Table 4. Percentage composition of saturates, aromatics, resins, asphaltenes and residue for modifiers.

Sample ID	Saturates (%)	Aromatics (%)	Resins (%)	Asphaltenes (%)	Residue (%)	Total (%)
A	8.3	19.7	62.5	4.6	4.9	100.0
C	0.0	0.0	70.1	23.3	6.6	100.0
D	1.2	10.5	78.5	7.5	2.3	100.0
E	0.0	0.1	83.4	13.0	3.5	100.0
K	59.3	0.4	10.3	4.0	26.0	100.0

Note: Results for modifier G cannot be determined using TLC-FID.

Modifier G possesses unique characteristics which are different from other modifiers. It is a water-soluble modifier and did not dissolve in the solvent used for other modifiers. Modifier G is also toluene insoluble which suggests there is no presence of asphaltic materials. Hence, the SARA approach seems inappropriate to characterize the chemical composition of modifier G. Furthermore, solubility of G in water can have moisture durability issues in AC designed from this modified binder which is not in the scope of this study. Therefore, AC properties to evaluate the effect of moisture should be investigated.

3.2. Rheological Characterization of Modified Binders

3.2.1. Superpave Grading

Base binder, S1 (PG 64-22), was modified to PG 58-28 using modifiers provided in Table 5. All the modifiers were able to convert the base binder to the acceptable limits of PG 58-28 except modifier K (ReOB). Figure 4a shows the continuous PG for base binder, unmodified PG 58-28 (S5), and modified binders. The selection of modifier dosage was based on (i) achieving similar high temperature true-grades (0.6 °C standard deviation) and (ii) obtaining similar true-grade results to S5's (±1.2 °C).

Table 5. Continuous and Superpave Performance Grade (PG) of base binder, unmodified binder and modified binders.

Binder ID	High PG	Low PG	Continuous PG	Superpave PG
S1	66.4	−23.7	66.4-23.7	64-22
S5	61.1	−29.5	61.1-29.5	58-28
S1-A-3.5	61.6	−30.0	61.6-30.0	58-28
S1-C-3.1	61.4	−30.1	61.4-30.1	58-28
S1-D-3.1	62.1	−28.3	62.1-28.3	58-28
S1-E-3.1	62.3	−30.7	62.3-30.7	58-28
S1-G-6.5	60.8	−28.5	60.8-28.5	58-28
S1-K-10	59.4	−27.3	59.4-27.3	58-22

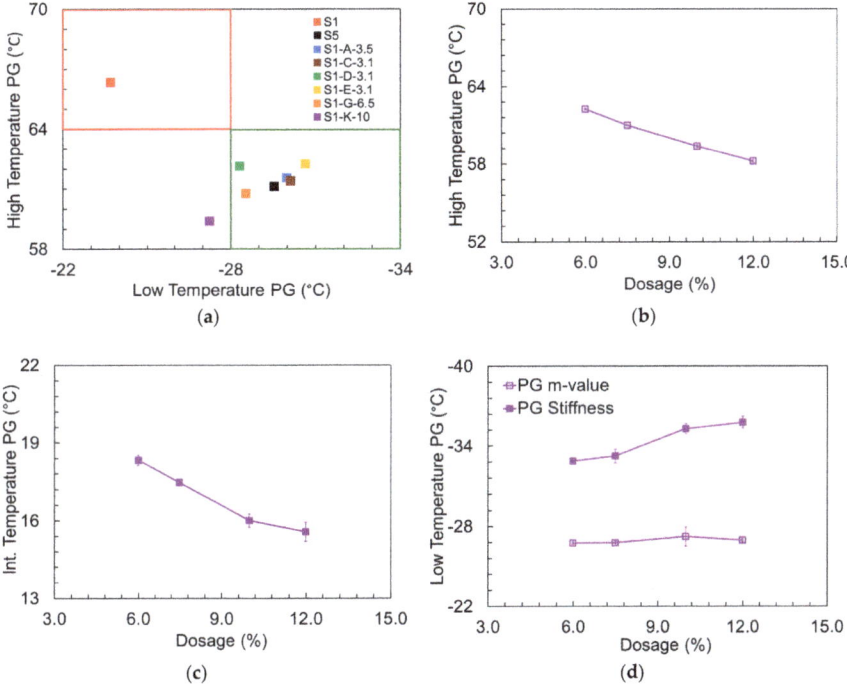

Figure 4. (a) Continuous PG of binders; for modifier K, effect of dosage on (b) High-temperature PG; (c) intermediate-temperature PG; and (d) Low-temperature $PG_{Stiffness}$ and $PG_{m\text{-value}}$.

For modifier K, none of the dosages from 6% to 12% were able to produce a PG 58-28 binder. It was observed that increase in the addition of modifier K had a softening effect on high and intermediate temperature grades (Figure 4b,c). In case of low temperature PG, the stiffness was reduced with increasing modifier dosage whereas m-value was not affected (as shown in Figure 4d). Moreover, the low temperature PG for K was controlled by the m-value which failed to meet required criteria. The 10% dosage for K was chosen to obtain the continuous PG closest to PG 58-28.

3.2.2. Delta T_c (ΔT_c)

ΔT_c for PAV, 2PAV, and 3PAV conditions was computed for all binders (Figure 5). The modification to S1 increased the ΔT_c values, at all aging conditions, for all modified binders except when modifier K (ReOB) was used. A higher, or less negative, ΔT_c can be related to better resistance to low-temperature cracking. Departments of Transportation in Maryland, Kansas, Pennsylvania, New York, New Jersey, Delaware, and Vermont restrict ΔT_c to be greater than −5 °C after 40-h PAV conditioning while some other states use a limit for ΔT_c after 20-h PAV [20].

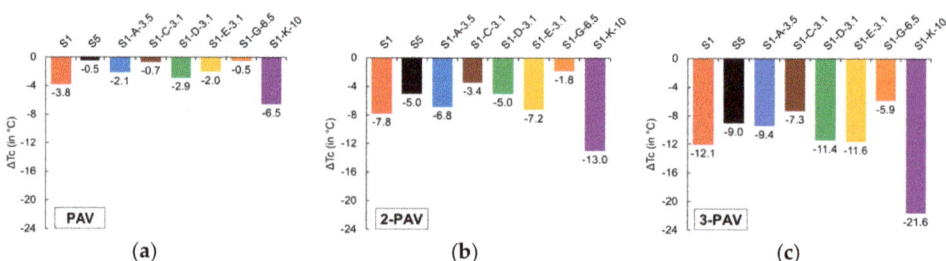

Figure 5. ΔT_c values of different modified binders, base binder and unmodified binder for aging conditions (**a**) pressure aging vessel (PAV); (**b**) 2PAV; and (**c**) 3PAV.

On comparing unmodified binder S5 to G and C modified binders, the later ones showed similar ΔT_c in PAV aging condition. However, with prolonged aging (2PAV and 3PAV), G-modified binder outperformed S5, followed by C-modified binder. MABs with modifiers A, D, E, and K had lower ΔT_c than S5. It is widely accepted that the presence of ReOB increases low-temperature cracking susceptibility of AB [5–9], which is also observed in this study. K-modified binder has the lowest ΔT_c at all aging conditions.

It is important to note that with aging, the effect on ΔT_c values are predominantly driven by the m-value criterion. In all cases, ΔT_c becomes more negative with aging, indicating more pronounced loss of relaxation (reduction of m-value) than stiffening of the material. Stiffnesses were not as greatly affected by aging.

The ΔT_c parameter suggests that modified binders may have similar long-term cracking resistance with the unmodified S5. On the other hand, K-modified binder (ReOB) has the lowest ΔT_c, which is well beyond the acceptable thresholds suggested in the literature [20].

3.2.3. Frequency Sweep Test

In this section, complex modulus master curves, black space diagrams, and rheological parameter, GR, are presented for, RTFO, PAV, 2PAV, and 3PAV aging conditions.

Complex Shear Modulus Master Curves

Complex modulus master curves were determined at a reference temperature of 15 °C. Figure 6a–d show the progression of complex modulus at different aging conditions, for limited low reduced-frequency range. This range was selected to represent low-temperature (non-load associated) cracking conditions. Following are the observations from Figure 6.

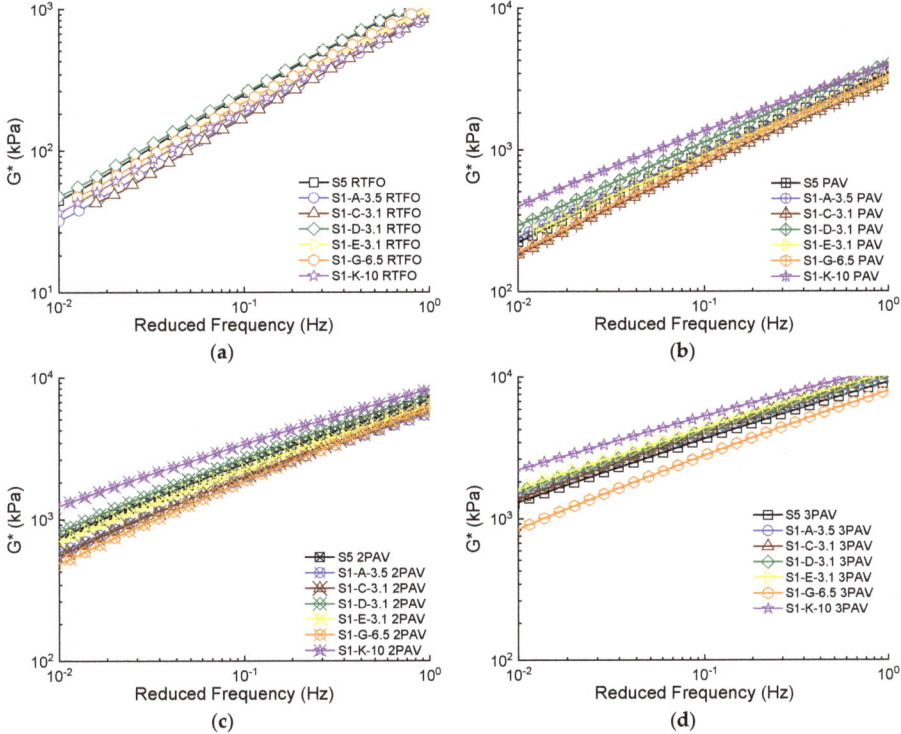

Figure 6. Complex shear modulus curves at reference temperature of 15 °C for aging conditions (**a**) Rolling Thin-Film Oven (RTFO); (**b**) PAV; (**c**) 2PAV; and (**d**) 3PAV.

1. In Figure 6a, after RTFO aging, modified binders are difficult to differentiate based on their shear modulus values. All the binders have similar master curves representing similar stiffnesses.
2. As the aging progressed to PAV condition, modified binder containing K separates from the rest of the binders at the lower frequency range (Figure 6b) representing a stiffer behavior. Other binders (A, C, D, E, G, and S5) show similar behavior.
3. Upon further aging (2PAV), modified binder containing G slightly separates from the rest of the binders towards lower moduli (Figure 6c), showing a softer behavior. K is again the stiffest and separates from the rest (A, C, D, E, and S5) in this condition.
4. At 3PAV, modified binders containing G and K have clear distinction in moduli, G being the softest and K being the stiffest. All other binders have similar variation in the complex shear modulus as shown in Figure 6d.

Black Space Diagram

Black space diagrams were plotted for RTFO, PAV, 2PAV, and 3PAV aging conditions. A black space diagram shows the variation of complex shear modulus (G*) with phase angle (δ). A reference temperature of 15 °C was used to obtain these curves. A higher δ at a fixed temperature and G* represents a material more prone to flow in a viscous manner and a lower δ represents a more elastic behavior [42]. Materials with higher δ are less likely to crack in a brittle way. In other words, a lower δ indicates that more energy would be stored, and faster accumulation of stress would be observed during repeated deformation [3]. A good correlation has been demonstrated between isothermal phase angle (especially at 50 °C) and test-pavement cracking severity [43]. Phase angle (at G* = 8967 kPa) has been found to be more repeatable measurement than G* and capable of identifying phase-incompatible asphalt binders [3].

Figure 7a–d show complete black space diagrams for specific aging conditions, whereas the inset figure highlights the effect on δ for G* ranging from 10 MPa to 11 MPa. The motivation for the selection of this G* range was the intermediate-temperature PG criteria [3]. However, a different range of G* selection does not affect the trends in δ and could represent other loading conditions in the field. Some observations from the black space diagrams are the following:

1. For the selected range of G* at RTFO (Inset Figure 7a) aged condition, K-modified binder has the lowest δ with clear distinction compared to other binders. The remaining binders have higher δ values than K-modified binder and are close to each other. C-modified binder has highest δ.
2. As the aging progresses to PAV aging, the separation in δ can be distinguished clearly. K-modified binder is separated from rest of the binders with lower δ (Inset Figure 7b), while G-modified binder has the highest δ followed by C, S5, E, D, and A, respectively (close to each other).
3. Upon further aging (2PAV and 3PAV), the trends for highest and lowest δ are again similar to that of PAV aging. However, the separation in δ became more evident (Inset Figure 7c,d). At 3PAV, K-modified binder has the lowest δ, followed by E, D, S5, C, and G. The δ values for G-modified binder suggest a greater viscous component in the complex shear modulus, indicating a flowing behavior, thus possibly lower cracking susceptibility.

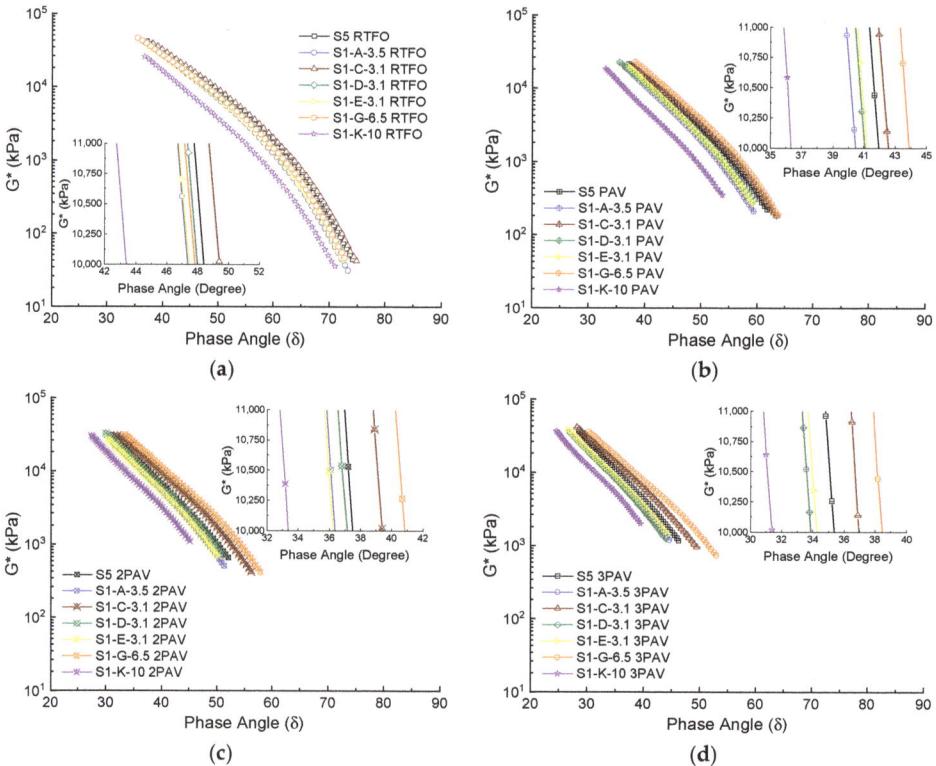

Figure 7. Black space diagrams at reference temperature of 15 °C for aging conditions (**a**) RTFO; (**b**) PAV; (**c**) 2PAV; and (**d**) 3PAV.

Glover–Rowe (GR) Parameter

GR values are indicated in the black space diagrams, shown in Figure 8, for all aging conditions. Modified binders containing A, D, and E show similar characteristics to unmodified binder (S5) at all aging conditions (Figure 8a). Differently, in case of modifier C, higher resistance to low-temperature cracking was observed.

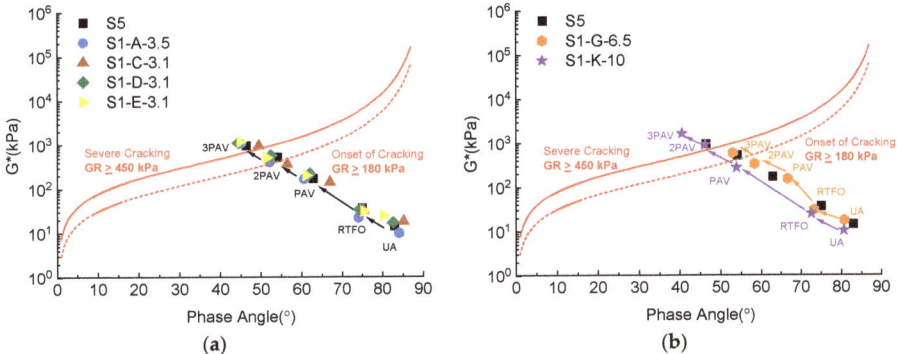

Figure 8. Black space diagrams showing G* vs phase angle (δ) at reference temperature of 15 °C and reduced frequency of 0.005 rad/s for all aging conditions for (**a**) modifiers A, C, D, and E compared to unmodified binder S5; and (**b**) modifiers G and K compared to unmodified binder S5.

A GR of 180 kPa is a criterion for damage onset for non-load-associated low-temperature cracking (shown in the red-dotted line in Figure 8), while 450 kPa is indicative for severe cracking (shown in the red-solid line in Figure 8) [23]. As per these thresholds, damage onset initiates around 2PAV while severe damage takes place at around 3PAV condition for S5 and binders containing modifiers A, D, and E.

In Figure 8b, for G-modified binder the evolution of GR parameter suggests higher resistance to aging compared to S5. The GR value for 3PAV of G is similar to that of 2PAV for S5. In contrast, GR for K-modified binder increases at a much faster rate than S5, suggesting potential early damage. The GR values of PAV and 2PAV for K are similar to 2PAV and 3PAV of S5, respectively. Figure 8a,b show that G-modified binder does not reach onset of severe cracking criterion after 3PAV while K-modified binder surpasses the criterion after 2PAV.

4. Discussion

4.1. Summary of Modifier Chemistry

The Elemental analysis showed that modifiers G and C had higher nitrogen content (9% and 3.5%, respectively), while E and K had some sulfur (0.33% and 0.98%, respectively) compared to other modifiers. Modifier G has significantly low carbon (47.9%) and high oxygen (33.1%) content.

FTIR spectra further validated the presence of nitrogen in modifiers C and G with peaks in the ranges of 1000–1250 cm^{-1} and 3100–3500 cm^{-1} which are characteristic of C-N stretching and N-H stretching (from secondary amines), respectively. A distinct peak at 2860 cm^{-1} and a small peak at 1650 cm^{-1} further validate the presence of nitrogen, as these peaks are representative of amine salt and amides, respectively. Modifiers A, C, D, and E show the presence of carbonyl functionality, which was observed from the carbonyl peaks at 1742 cm^{-1}.

Molecular weight analysis showed that modifiers A, C, D, and E have average molecular weights in a similar range, modifier G has the lowest weight and K has the highest. In addition, modifier K possesses significantly high PDI indicating a wide variation of molecular species presence. Modifiers, however, have different molecular size distributions, with some modifiers having multiple peaks (A, C, D, and K) and hence, are composed of distinct molecules. On the other hand, others (E and G) had single peak which means they are composed of single molecular size.

Modifier G has a distinct chemical composition compared to other modifiers. It is significantly high on nitrogen and oxygen and relatively low on carbon compared to others. The presence of nitrogen was validated by the FTIR spectra. Peaks corresponding to amine salt and primary and secondary amines were observed. It was found that the molecular weight of G was the lowest single peak distribution. Additionally, modifier C showed relatively higher content of elemental nitrogen in the form of secondary amines which was verified by the FTIR spectrum.

SARA analysis of G was inconclusive and additional characterization with a different method is required to evaluate the chemical characteristics of modifier G. Moreover, its solubility in water requires additional investigation of the AC's susceptibility to moisture. Modifiers C and E are largely composed of resins and asphaltenes with limited or no saturates and aromatics. While A and D have some aromatics and asphaltenes along with a majority of resins. Modifier K has large amount of insoluble residue with high saturate content and limited resins with traces of aromatics.

Furthermore, modifier A shows similar chemical functional groups, molecular weight and molecular weight distribution to modifier D, which indicates that modifier A may belong to bio-oil category.

4.2. Summary of Modified Asphalt Binder Rheology

Modifiers' dosages were selected to meet PG 58-28 with true grades close to each other for all MABs to ensure reasonable comparison of rheological properties, except for binder modified with K (ReOB). It was observed that increasing the dosage of K increased high PG, decreased intermediate PG, and decreased low PG stiffness. However, there was no improvement in the relaxation properties of the modified binder with increasing amount of modifier K in the blend. Therefore, dosage for modifier K resulting in a continuous PG closest to PG 58-28 was selected for further investigation. The selected dosage varied from one modifier to the other. This might be one of the reasons for the observed differences in rheology. For instance, increasing the dosage of modifier C can result in a similar performance as that of AB modified with modifier G; but would result in a different Superpave continuous-PG. The focus of the study was to compare binders with similar Superpave characteristics. Dosage variation/optimization is not within the scope of this study, but appears to be a promising research path.

ΔT_c was determined to evaluate the low temperature cracking susceptibility of modified binders for PAV, 2PAV, and 3PAV aging conditions. Relatively high ΔT_c indicates better relaxation properties at low temperatures, which results in better resistance to cracking. Modification of S1 improved the ΔT_c for all MABs except K. Significant improvements were observed when modifiers G and C were used, which even showed better relaxation properties than unmodified binder (S5). G-modified binder has the highest ΔT_c for 2PAV and 3PAV conditions, followed by binder modified with C. Differently, K-modified binder has the lowest ΔT_c values in all aging conditions. MABs containing A, D, and E have ΔT_c values close to S5 only after 2PAV and 3PAV aging conditions.

Complex shear modulus master curves show that modulus consistently increased for all aged binders. K-modified binder showed distinctly stiff behavior at PAV that was also observed in 2PAV and 3PAV conditions. Other MABs stiffness trends shifted with aging conditions. At UA condition, G-modified binder was the stiffest and after 2PAV and 3PAV, it was the softest binder; which is desired. However, mechanisms of change in modulus for G-modified binder after aging need to be investigated. Modifier G, as discussed before, has distinctive characteristics and needs to be explored with additional testing. Aging after 2PAV and 3PAV, other modifiers have master curves closer to unmodified binder (S5).

The black space diagram was used to evaluate the impact of aging on phase angle (δ). The δ for selected range of G* shows similar trends after PAV, 2PAV and 3PAV aging. The differences in δ of

MABs become more distinct with aging but are always noticeable, which makes δ at a certain G* a useful parameter to distinguish MABs. Note that this might not be the case when polymers are in the blend [44]. As mentioned earlier, a higher δ at a certain G* indicates that the material is less prone to cracking in a brittle way at service conditions. Again, MAB containing G has the highest phase angles, followed by C, S5, E, D, and A, while K has the lowest. In addition, the evolution of GR parameter also suggests that G-modified binder is the most resistant to aging while K is the least.

Based on rheological testing, G-modified binder is least susceptible to cracking followed by C-modified binder whereas K-modified binder is the most susceptible. MABs modified with A, D and E show similar rheological characteristics to S5.

4.3. Relationship between Modifier Chemistry and Binder Rheology

Modifier's chemical make-up contributed significantly to the long-term rheological response of MABs. Nitrogen-based compounds are known for their antioxidant properties [45]. The presence of higher nitrogen content was validated with elemental analyses and FTIR spectra and its impact was observed in the change of frequency sweep measurements as aging progressed. Modifier C, containing 3.5% nitrogen, shows similar or better crack resisting properties than the unmodified product (S5) at PAV, 2PAV, and 3PAV conditions. Superior rheological properties of G-modified binder can be attributed to the presence of high nitrogen content (9.0% in modifier G), which is composed of nitrogen-based compounds like amines. This validates the impact of antioxidants on resisting binder aging, and hence, reducing cracking susceptibility.

On the other hand, sulfur presence in modifiers E (0.33%) and K (0.98%) is accompanied by lower expected performance based on the reported rheological parameters. Excessive sulfur content (>4%) in binders can cause increased oxidation due to the formation of additional sulfoxides causing embrittlement in binders [46].

Lower molecular weight of modifiers could be promoting phase compatibility. Rheological test results and M_w distinguish three groups: MABs containing A, D, and E have similar characteristics, MAB containing K has lower expected performance and highest M_w, and MABs containing C and G have higher expected performance and lower M_w.

The rheological parameters: ΔT_c, GR, and phase angle from black space diagram have consistent trends among all modifiers and are able to distinguish MABs based on their expected cracking performance.

5. Summary and Findings

The focus of this paper is to evaluate the impact of modifiers' chemical properties on the rheological properties of respective modified binders. Binders blended with various types of modifiers, intended to soften (reduce) the grade of an unmodified binder, were tested at various aging conditions (unaged, RTFO, PAV, 2PAV, and 3PAV). Performance progression indicators were used to predict their long-term performance. Low-temperature cracking susceptibility was assessed using GR and ΔT_c, and intermediate-temperature cracking susceptibility was assessed using black space diagrams. Chemical characteristics of modifiers were evaluated using elemental analysis, FTIR, GPC, and TLC-FID. The results show that modifier chemistry impacts modified binder performance. The presence of certain elements, chemical functional groups and molecular size can affect the rheological properties of the binder. Following are the findings of this study:

1. Nitrogen-rich modified binders appear to have superior rheological properties. They have higher ΔT_c, higher phase angles, and lower GR for 2PAV and 3PAV conditions. Hence, it can be assumed that the presence of nitrogen would boost anti-oxidizing properties and reduce susceptibility to cracking. Further research is needed to validate this hypothesis.
2. Sulfur presence may have a detrimental impact on modified binder performance.

3. Lower molecular size/weight of modifiers appears to be related to better cracking resistance potential of modified asphalt binders. Further research towards validating this idea is encouraged.
4. Modifiers A and D have similar chemical characteristics and molecular weight distributions. This suggests that modifier A may have a bio-based origin.
5. The methodology employed in the SARA analysis may not be able to characterize some modifiers. An alternative test method should be considered for some modifiers.
6. Phase angle parameter in conjunction with complex modulus (G*) was shown to be sensitive to laboratory aging of modified and unmodified binders. Therefore, such a parameter has potential to identify phase-incompatible asphalt binders, is able to distinguish potential mechanical behavior, and could be associated with field performance.

In conclusion, modifier chemistry was shown to have a relationship with rheological behavior of modified binders. The experimental program presented in this paper can be used to choose modifiers that may control cracking development and could also be used as guidance to engineer asphalt binder modifiers.

Author Contributions: All authors have read and agree to the published version of the manuscript. Conceptualization, B.K.S., H.O. and I.L.A.-Q.; Methodology, B.K.S., H.O. and I.L.A.-Q.; Validation, P.S. and J.J.G.M.; Formal analysis, P.S. and J.J.G.M.; Investigation, B.K.S., H.O. and I.L.A.-Q., J.J.G.M. and P.S.; Resources, I.L.A.-Q., H.O. and B.K.S.; Data curation, P.S. and J.J.G.M.; Writing—original draft preparation, P.S. and J.J.G.M.; Writing—Review and editing, I.L.A.-Q., B.K.S., H.O.; Visualization, P.S. and J.J.G.M.; Supervision, B.K.S., H.O. and I.L.A.-Q.; Project administration, B.K.S., H.O., and I.L.A.-Q.; Funding acquisition, B.K.S., H.O. and I.L.A.-Q.

Funding: This research was funded by the Illinois Department of Transportation, under the project number ICT R27-196 HS.

Acknowledgments: This publication is based on the results of "ICT-R27-196 HS: Rheology-Chemical Based Procedure to Evaluate Additives/Modifiers used in Asphalt Binders for Performance Enhancements (Phase 2)." ICT-R27-196 HS is currently conducted in cooperation with the Illinois Center for Transportation (ICT); Illinois Sustainability Technology Center (ISTC); the Illinois Department of Transportation (IDOT); and the U.S. Department of Transportation, Federal Highway Administration. Special thanks to ICT and ISTC students and research staff, Kirtika Kohli, Uthman Mohammed Ali, Greg Renshaw, and Marc Killion, for their input and support during this study. The contributions of the technical review committee are acknowledged; special thanks to Kelly Morse, Jim Trepanier, Ronald Price, Clay Snyder and Brian Hill. We also thank binder and modifiers suppliers for providing the requested quantities of the samples. The contents of this paper reflect the view of the authors, who are responsible for the facts and the accuracy of the data presented herein. The contents do not necessarily reflect the official views or policies of ICT, ISTC or IDOT. This paper does not constitute a standard, specification, or regulation.

Conflicts of Interest: The authors declare no conflict of interest. The funders had no role in the design of the study; in the collection, analyses, or interpretation of data; in the writing of the manuscript, or in the decision to publish the results.

References

1. Williams, B.A.; Willis, R.J.; Ross, C.T. *Annual Asphalt Pavement Industry Survey on Recycled Materials and Warm-Mix Asphalt Usage 2018*; National Asphalt Pavement Association: Greenbelt, MD, USA, 2018.
2. Van Krevelen, D.W.; Te Nijenhuis, K. Chemical Degradation. In *Properties of Polymers*; Elsevier: Amsterdam, The Netherlands, 2009.
3. Kriz, P.; Noel, J.; Quddus, M.; Maria, S. Rheological Properties of Phase-Incompatible Bituminous Binders. In Proceedings of the 56th Peterson Asphalt Research Conference, Laramie, WY, USA, 14–17 July 2019.
4. *Asphalt Institute and Eurobitume The Bitumen Industry-A Global Perspective: Production, Chemistry, Use, Specification and Occupational Exposure*; Asphalt Institute and Eurobitume: Lexington, KY, USA, 2015.
5. Ozer, H.; Renshaw, G.; Hasiba, K.; Al-Qadi, I.L. *Evaluation of the Impacts of Re-Refined Engine Oil Bottoms (ReOB) on Performance Graded Asphalt Binders and Asphalt Mixtures, Report No. FHWA-ICT-16-006*; Illinois Center for Transportation: Rantoul, IL, USA, 2016.
6. Rose, A.A.; Lenz, I.R.; Than, C.T.; Glover, C.J. Investigation of the effects of recycled engine oil bottoms on asphalt field performance following an oxidation modeling approach. *Pet. Sci. Technol.* **2016**, *34*, 1768–1776. [CrossRef]

7. Jia, X.; Huang, B.; Moore, J.A.; Zhao, S. Influence of waste engine oil on asphalt mixtures containing reclaimed asphalt pavement. *J. Mater. Civ. Eng.* **2015**, *27*, 1–9. [CrossRef]
8. Reinke, G.; Hanz, A.; Anderson, R.M.; Ryan, M.; Engber, S.; Herlitzka, D. Impact of re-refined engine oil bottoms on binder properties and mix performance on two pavements in Minnesota. In Proceedings of the E&E Congress 2016, 6th Eurasphalt & Eurobitume Congress, Prague, Czech Republic, 1–3 June 2016.
9. Golalipour, A.; Bahia, H. Investigation of Effect of Bio-based and Re-refined Used Oil Modifiers on Asphalt Binder's Performance and Properties. *Transp. Res. Board* **2016**, 1–14.
10. Fini, E.H.; Oldham, D.; Buabeng, F.S.; Nezhad, S.H. Investigating the Aging Susceptibility of Bio-Modified Asphalts. *Airf. Highw. Pavement* **2015**, 62–73. [CrossRef]
11. Borghi, A.; del Barco Carrión, A.J.; Lo Presti, D.; Giustozzi, F. Effects of Laboratory Aging on Properties of Biorejuvenated Asphalt Binders. *J. Mater. Civ. Eng.* **2017**, *29*, 1–13. [CrossRef]
12. Al-Qadi, I.L.; Wu, S.; Lippert, D.L.; Ozer, H.; Barry, M.K.; Safl, F.R. Impact of high recycled mixes on HMA overlay crack development rate. *Asph. Paving Technol. Assoc. Asph. Paving Technol. Tech. Sess.* **2017**, *86*, 427–447.
13. Al-Qadi, I.L.; Ozer, H.; Zhu, Z.; Singhvi, P.; Ali, M.U.; Sawalha, M.; Luque, A.E.F.; Mainieri, J.J.G.; Zehr, T.G. *Development of Long-Term Aging Protocol for Implementation of the Illinois Flexibility Index Test (I-FIT)*; Illinois Center for Transportation: Rantoul, IL, USA, 2019.
14. Doll, B.; Ozer, H.; Rivera-Perez, J.J.; Al-Qadi, I.L.; Lambros, J. Investigation of viscoelastic fracture fields in asphalt mixtures using digital image correlation. *Int. J. Fract.* **2017**, *205*, 37–56. [CrossRef]
15. Ozer, H.; Al-Qadi, I.L.; Lambros, J.; El-Khatib, A.; Singhvi, P.; Doll, B. Development of the fracture-based flexibility index for asphalt concrete cracking potential using modified semi-circle bending test parameters. *Constr. Build. Mater.* **2016**, *115*, 390–401. [CrossRef]
16. Ozer, H.; Al-Qadi, I.L.; Singhvi, P.; Khan, T.; Rivera-Perez, J.; El-Khatib, A. Fracture Characterization of Asphalt Mixtures with High Recycled Content Using Illinois Semicircular Bending Test Method and Flexibility Index. *Transp. Res. Rec.* **2016**, *2575*, 130–137. [CrossRef]
17. Wu, S.; Al-Qadi, I.L.; Lippert, D.L.; Ozer, H.; Luque, A.F.E.; Safi, F.R. Early-age performance characterization of hot-mix asphalt overlay with varying amounts of asphalt binder replacement. *Constr. Build. Mater.* **2017**, *153*, 294–306. [CrossRef]
18. Wu, S.; Lippert, D.L.; Al-Qadi, I.L.; Ozer, H.; Said, I.M.; Barry, M.K. Impact of High Asphalt Binder Replacement on Level Binder Properties for Controlling Reflective Cracking. *Transp. Res. Rec. J. Transp. Res. Board* **2017**, *2630*, 118–127. [CrossRef]
19. Anderson, M.R.; King, G.N.; Hanson, D.I.; Blankenship, P.B. Evaluation of the Relationship between Asphalt Binder Properties and Non-Load Related Cracking. *J. Assoc. Asph. Paving Technol.* **2011**, *80*, 615–662.
20. Asphalt Institute. *Use of the Delta Tc Parameter to Characterize Asphalt Binder Behavior IS-240*; Asphalt Institute: Lexington, KY, USA, 2019.
21. Kandhal, P. *Low-Temperature Ductility in Relation to Pavement Performance*; ASTM International: West Conshohocken, PA, USA, 1977.
22. Glover, C.J.; Davison, R.R.; Domke, C.H.; Ruan, Y.; Juristyarini, P.; Knorr, D.B.; Jung, S.H. *Development of a New Method for Assessing Asphalt Binder Durability with Field Validation*; Texas Transportation Institute: College Station, TX, USA, 2005.
23. Rowe, G. Analysis of SHRP core asphalts—New 2013/14 test results. In Proceedings of the Binder Expert Task Group Meeting, San Antonio, TX, USA, 1–2 April 2014.
24. Hintz, C.; Velasquez, R.; Johnson, C.; Bahia, H. Modification and Validation of Linear Amplitude Sweep Test for Binder Fatigue Specification. *Transp. Res. Rec.* **2011**, *2207*, 99–106. [CrossRef]
25. Elwardany, M.D.; Rad, F.Y.; Castorena, C.; Richard Kim, Y. Climate-, Depth-, and Time-based Laboratory Aging Procedure for Asphalt Mixtures. *Asph. Paving Technol. Assoc. Asph. Paving Technol. Tech. Sess.* **2018**, *87*, 467–511. [CrossRef]
26. D'Angelo, J.A. Asphalt Binders and aging 20hr or 40hr PAV. In Proceedings of the Binder Expert Task Group Meeting, Fall River, MA, USA, 9–10 May 2018.
27. Singhvi, P.; Karakas, A.; Ozer, H.; Al-Qadi, I.L.; Hossain, K. Impact of Asphalt Modifier Dosage on Modified Binder Rheology and Chemistry with Long-Term Aging. In Proceedings of the Airfield and Highway Pavements 2019, Chicago, IL, USA, 21–24 July 2019.

28. Petersen, J.C. Chemical Composition of Asphalt As Related To Asphalt Durability: State of the Art. *Transp. Res. Rec.* **1984**, *999*, 13–30.
29. Sharma, B.K.; Ma, J.; Kunwar, B.; Singhvi, P.; Ozer, H.; Rajagopalan, N. *Modeling the Performance Properties of Ras and Rap Blended Asphalt Mixes Using Chemical Compositional Information*; Illinois Center for Transportation: Rantoul, IL, USA, 2017.
30. Sharma, B.K.; Singhvi, P.; Kunwar, B.; Kohli, K.; Ma, J.; Ozer, H.; Al-Qadi, I.L. Aging and recycling effect on the relationship between the chemistry and rheological parameters for binders. *Fuel* (under review).
31. Sáda Costa, M.; Farcas, F.; Santos, L.F.; Eusébio, M.I.; Diogo, A.C. Chemical and Thermal Characterization of Road Bitumen Ageing. *Mater. Sci. Forum* **2010**, *636–637*, 273–279.
32. Xiang, L.; Cheng, J.; Kang, S. Thermal oxidative aging mechanism of crumb rubber/SBS composite modified asphalt. *Constr. Build. Mater.* **2015**, *75*, 169–175. [CrossRef]
33. Abbas, A.R.; Mannan, U.A.; Dessouky, S. Effect of recycled asphalt shingles on physical and chemical properties of virgin asphalt binders. *Constr. Build. Mater.* **2013**, *45*, 162–172. [CrossRef]
34. Sun, Z.; Yi, J.; Huang, Y.; Feng, D.; Guo, C. Properties of asphalt binder modified by bio-oil derived from waste cooking oil. *Constr. Build. Mater.* **2016**, *102*, 496–504. [CrossRef]
35. Poulikakos, L.D.; Santos, S.; dos Bueno, M.; Kuentzel, S.; Hugener, M.; Partl, M.N. Influence of short and long term aging on chemical, microstructural and macro-mechanical properties of recycled asphalt mixtures. *Constr. Build. Mater.* **2014**, *51*, 414–423. [CrossRef]
36. Robertson, R.E. *Chemical Properties of Asphalts and Their Relationship to Pavement Performance*; Western Research Institute: Laramie, WY, USA, 1991.
37. Liu, G.; Nielsen, E.; Komacka, J.; Leegwater, G.; van de Ven, M. Influence of soft bitumens on the chemical and rheological properties of reclaimed polymer-modified binders from the "old" surface-layer asphalt. *Constr. Build. Mater.* **2015**, *79*, 129–135. [CrossRef]
38. Branthaver, J.F.; Petersen, J.C.; Robertson, R.E.; Duvall, J.J.; Kim, S.S.; Harnsberger, P.M.; Mill, T.; Ensley, E.K.; Barbour, F.A.; Schabron, J.F. Binder Charact. *Eval* **1993**, *2*, 1–467.
39. Yu, X.; Zaumanis, M.; dos Santos, S.; Poulikakos, L.D. Rheological, microscopic, and chemical characterization of the rejuvenating effect on asphalt binders. *Fuel* **2014**, *135*, 162–171. [CrossRef]
40. Huang, S.-C.; Pauli, A.T.; Grimes, R.W.; Turner, F. Ageing characteristics of RAP binder blends – what types of RAP binders are suitable for multiple recycling? *Road Mater. Pavement Des.* **2014**, *15*, 113–145. [CrossRef]
41. Gong, M.; Yang, J.; Zhang, J.; Zhu, H.; Tong, T. Physical–chemical properties of aged asphalt rejuvenated by bio-oil derived from biodiesel residue. *Constr. Build. Mater.* **2016**, *105*, 35–45. [CrossRef]
42. Anderson, D.A.; Chris, D.W.; Bahia Hussain, U.; Dongre, R.; Sharma, M.G.; Antle Charles, E.; Button, J. Binder Characterization and Evaluation Volume 3. *Phys. Charact.* **1994**, *3*, 1–475.
43. Pauli, A.; Farrar, M.; Huang, S.-C. Characterization of Pavement Performance Based on Field Validation Test Site Data Interpreted by an Asphalt Composition Model of Binder Oxidation. *Transp. Res. Circ.* **2018**, *E-C234*, 34–62.
44. Airey, G.D. Use of Black Diagrams to Identify Inconsistencies in Rheological Data. *Road Mater. Pavement Des.* **2002**, *3*, 403–424. [CrossRef]
45. Speight, J.G. The Chemistry and Technology of Petroleum. In *The Chemistry and Technology of Petroleum*; CRC Press: Boca Raton, FL, USA, 2014.
46. Peterson, J.C. A Dual, Sequential Mechanism for the Oxidation of Petroleum Asphalts. *Pet. Sci. Technol.* **1998**, *16*, 1023–1059. [CrossRef]

© 2020 by the authors. Licensee MDPI, Basel, Switzerland. This article is an open access article distributed under the terms and conditions of the Creative Commons Attribution (CC BY) license (http://creativecommons.org/licenses/by/4.0/).

Article

An Assessment of Moisture Susceptibility and Ageing Effect on Nanoclay-Modified AC Mixtures Containing Flakes of Plastic Film Collected as Urban Waste

Arminda Almeida [1,2,*], João Crucho [3], César Abreu [4] and Luís Picado-Santos [3,5]

1. Department of Civil Engineering, University of Coimbra, 3030-788 Coimbra, Portugal
2. CITTA-Research Centre for Territory, Transports and Environment, 4200-465 Oporto, Portugal
3. CERIS- Civil Engineering Research and Innovation for Sustainability, Instituto Superior Técnico, Universidade de Lisboa, 1049-001 Lisbon, Portugal
4. ACIV- Associação para o Desenvolvimento da Engenharia Civil, Department of Civil Engineering, University of Coimbra, 3030-788 Coimbra, Portugal
5. Instituto Superior Técnico, Universidade de Lisboa, 1049-001 Lisbon, Portugal
* Correspondence: arminda@dec.uc.pt; Tel.: +351-239-797-104

Received: 10 June 2019; Accepted: 4 September 2019; Published: 7 September 2019

Abstract: In this research, the moisture susceptibility of a nanoclay-modified asphalt concrete (AC) mixture containing plastic film (in flakes) collected as urban waste was evaluated with specimens subjected to the tecnico accelerated ageing (TEAGE) procedure. The TEAGE procedure attempts to simulate—in a laboratory setting—the effect of field ageing by applying watering/drying cycles and ultraviolet radiation. For comparison purposes, three AC mixtures were considered, one for control, without plastic and nanoclay, a mixture with only plastic, and a mixture with both plastic and nanoclay. Furthermore, only half of the specimens were subjected to the ageing procedure. The plastic was added to the mixture using the dry process, and the nanoclay was blended with the bitumen before mixture preparation. The moisture susceptibility was evaluated, using a total of 48 Marshall specimens, by the indirect tensile strength ratio (ITSR). From the results of this study, the nanoclay-modified AC mixture containing plastic film presented slightly higher indirect tensile strength (ITS) values, lower moisture susceptibility, and enhanced ageing resistance. These slight improvements can be justified by the reduced air voids content of the samples and consequently they must be seen as conservative. Nevertheless, the modification of AC mixtures with flakes of plastic and nanoclay can be a viable solution for the recycling of plastic film collected as urban waste, being an eco-friendly alternative to disposal in landfills.

Keywords: ageing; plastic film; urban waste; nanoclay; moisture; indirect tensile strength

1. Introduction

Plastics cover a wide range of synthetic polymers, being used for the production of a very wide range of products that are extensively used in our everyday life. They are applied in many industry sectors, such as building and construction, electronics, automotive, agriculture, health, packaging, and energy [1], making the world plastic production increase year by year at a rate of about 4%, reaching almost 350 million tonnes in 2017 (65 million tonnes in Europe) [2]. After use, 42% of post-consumer waste plastics are incinerated and 27% are lost in landfills [1]. Only the remaining share (31%) is recycled. Packaging is the most demanded sector with 40% of all plastic production, 63% of all plastic waste, and 83% of all plastic that is recycled [3]. In terms of polymer, roughly 50% of plastic packaging waste from households is plastic film, mainly in the form of low-density polyethylene (LDPE) [3,4]. Plastic film includes all flexible packaging, such as grocery bags, food storage bags, product wraps, cling wrap, etc. They are mostly single-use plastic products.

Despite the benefits of plastics, the way some plastic products are produced, used, and discarded poses some problems. Most of them come from fossil sources being non-biodegradable, which leads to high contamination with ecological and health implications [5]. To minimize these problems, recycling rate targets have been imposed [6] and a transition to a circular economy has been fostered [7,8]. In a circular economy, plastic waste is considered as a resource capable of substituting virgin material.

As an alternative to virgin polymers, plastic waste has been used as additives to modify bitumen with the purpose of enhancing asphalt concrete (AC) mixtures with economic and environmental benefits [9,10]. The modification can be done either by mixing the additive previously with bitumen (wet process) [11] or by adding the additive directly to aggregates during the blending process (dry process) [9,12,13]. The dry process is simpler and more economical than the wet one [9], being considered by many researchers to be the best one [14,15]. The addition of waste thermoplastic polymers has been widely studied, e.g., polyethylene (PE) [15,16], polypropylene (PP) [11], polystyrene (PS) [17], polymerizing vinyl chloride (PVC) [10], polyethylene terephthalate (PET) [18], as well as combined polymers [9]. Regarding plastic film waste, studies have been conducted using plastic film from the agricultural sector [12,19] or focusing on specific plastic sources, like bags [20]. There appears to be a gap in the literature regarding the incorporation of plastic film waste (in flakes) collected as urban waste in AC mixtures, which represent a huge percentage of plastic packaging waste.

AC mixtures are subjected to traffic loads as well as climatic loads over their lifetime. In contrast to traffic loads, which have been investigated in several studies, the effect of climatic conditions (moisture, oxygen, heat, cold, ultraviolet (UV) light, freeze–thaw cycles, etc.) during prolonged periods is not well established, mainly when the AC mixtures have additives incorporated, even though they contribute considerably to the degradation of the pavement structures. Moisture susceptibility, defined as the loss of mechanical characteristics of materials resulting from the presence of water in AC mixtures, is commonly evaluated using the indirect tensile strength ratio (ITSR). This susceptibility is a complex process that depends on several factors, such as aggregate mineral composition, bitumen grade, bitumen-aggregate adhesion, and air voids (not only the content but also their distribution and connectivity) as well as the interaction between them [21–25]. In general, plastic waste additives increase moisture resistance of AC mixtures [16]. But, regrettably, this is not always the case [9]. The incorporation of additives might increase AC mixtures behaviour complexity. Diab, et al. [26] studied the moisture resistance (using ITSR) of polymer-modified AC mixtures (six different polymeric products were tested). They found that polymeric products, not only the type but also the content, have different moisture resistance. AC mixtures are subjected to environmental conditions as they age and consequently their properties evolve over time due to interaction of these factors. Age hardening and moisture damage have been pointed to as the primary factors affecting the durability of AC mixtures [27]. Nevertheless, even if there are more studies on moisture susceptibility, there are also studies considering moisture and ageing simultaneously [26,28–30], highlighting the importance to consider both.

The AC mixture ageing is generally divided into two phases; the phase of mixing, laying, and compaction, where the short-term ageing occurs; and, the pavement service life phase, where the long-term ageing occurs. To simulate ageing in the laboratory, the methods more frequently used are those described by the AASHTO standard practice R30-02 [31]. This standard proposes a method—the long-term oven ageing (LTOA)—to simulate the long-term ageing of the compacted asphalt mixture to an equivalent of 7–10 years of field service. On one hand, the LTOA method has several advantages, such as, it uses simple equipment and it is easy to implement and reproduce (consists of placing the compacted test specimens in a conditioning oven for 120 ± 0.5 h at a temperature of 85 ± 3 °C). On the other hand, it presents a few drawbacks, particularly, by not simulating most of the actions present during field ageing (e.g., solar radiation, moisture, and freeze-thaw), the field-ageing equivalency is difficult and not consensual. Smith Braden and Howard Isaac [32] matched the damage caused by laboratory conditioning protocols to the damage produced by exposure to non–load-associated environmental factors for up to 5 years in the southeast United States, concluding that the 7–10 years of field service simulation claimed by the AASHTO R30-02 are not realistic. It depends greatly on climate [29] and the procedure described in AASHTO R30-02 does not consider the moisture and UV radiation. Several authors are highlighting the importance of solar radiation, particularly the UV, in the ageing process [33–35]. As the UV rays possess high energy, they promote photodegradation mechanisms, for instance, they are able to break C=C bonds, and accelerate the degradation of polymers such as the styrene-butadiene-styrene (SBS) [36]. Thus, UV is an important action to reproduce when evaluating the ageing of a mixture to be applied in wearing course (surface layers) and when new additives/modifiers are involved. In order to account for these aspects, Crucho, Picado-Santos, Neves, Capitão, and Al-Qadi [29] developed a new accelerated ageing method for compacted bituminous mixtures, called tecnico accelerated ageing (TEAGE). It simulates the ageing of asphalt mixtures under specific environmental conditions by applying watering/drying cycles and UV radiation in equivalent levels to those observed in the field during a certain time. Those authors [29] compared the effect of TEAGE and R30-02 long-term ageing methods on an AC 14 gap-graded mixture by using stiffness, fatigue resistance, and indirect tensile strength (ITS) results. Compared to the R30-02 long-term ageing method, the ITS values of TEAGE aged specimens were higher, stiffness values were lower, as well as fatigue resistance values. However, the results are not comparable since the section of TEAGE aged specimens is not homogeneous, due to a differential ageing level in specimen depth (e.g., the top surface, with direct exposure to UV radiation suffered a higher ageing severity than the bottom surface). To consider that heterogeneity, the authors recovered the bitumen and carried out tests over it, whose results were used to predict stiffness through the depth of the specimen, concluding that the TEAGE method addresses in a more consistent way the ageing mechanisms that could be seen in the field.

The ageing effect has been addressed in several studies, e.g., [37–40]. Islam, et al. [41] studied, for the first time, ageing (long-term and short-term) effects using ITS tests, and found that for long-term oven-aged specimens, ITS increases with ageing, while for short-term oven-aged loose samples, ITS increases with the conditioning period until reaching a peak and then decreases.

On these grounds and taking into account that there is a need to use new materials to obtain both sustainable and high-performance pavements (to support an increase of traffic intensity, the presence of large and heavier trucks, and all the climatic agents), nanomaterials have attracted increasing attention in bitumen modification. They tend to improve AC mixture performance [42–46]. Regarding moisture, nanomaterials can work as antistrip additives, improving aggregates coating, and reducing thus moisture susceptibility [47–51]. Nanoclay has been used successfully [43], besides other properties, to improve moisture resistance as well as ageing resistance [50,52].

There are in the literature, studies encompassing moisture, ageing, polymers, and nanomaterials [26,50,52]. However, to date, few or no studies consider moisture, ageing, nanomaterials, and plastic waste simultaneously, i.e., a set of issues that can interact between themselves and may change AC mixture behaviour. This work attempted to contribute to that gap in literature by evaluating the moisture susceptibility of a nanoclay modified-AC mixture with flakes of plastic film collected as urban waste with and without ageing. The moisture susceptibility is evaluated by the ITSR value and the TEAGE method was used to simulate long-term ageing in a laboratory. For a better interpretation of the results, two additional AC mixtures are tested in parallel for comparison purposes, via a control AC mixture and a mixture with only plastic flakes. In addition, aged and unaged specimens are considered.

2. Materials and Methods

The materials and methods used in this study were selected to assess the moisture susceptibility and the ageing effect on nanoclay-modified AC mixtures containing flakes of plastic film collected as urban waste.

2.1. Materials

An asphalt mixture with a maximum aggregate size of 14 mm was selected (AC 14), which is used in surface layers. The materials used for the production of the AC mixtures are those described below.

2.1.1. Aggregates and Bitumen

The AC mixtures were produced with two types of aggregates (gneiss and limestone) and a conventional 35/50 paving grade bitumen supplied by Cepsa Portugal (located in Matosinhos, Portugal). For the mixture of aggregates the following aggregate fractions were selected: fraction 0/4 of crushed gneiss; fraction 4/8 of crushed gneiss; fraction 6/14 of crushed gneiss; and, limestone filler. The gradation limits used for the AC 14 mixture are the ones defined in the Portuguese road administration specifications [53]. Figure 1 presents the aggregate gradation as well as the gradation limits.

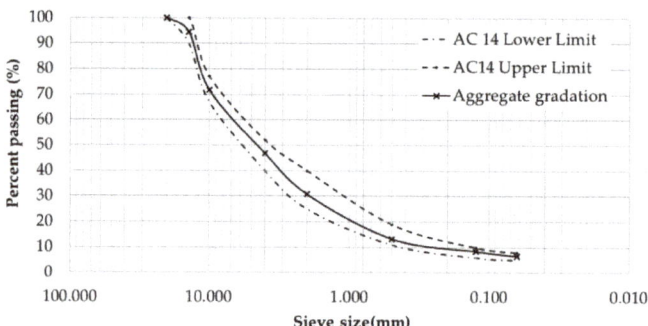

Figure 1. Aggregate gradation of the asphalt concrete with an aggregate size of 14 mm (AC 14) mixture.

2.1.2. Plastic

The flakes of plastic film collected as urban waste came from an LDPE plastic recycling plant (Ambiente-Recuperação de Materiais Plásticos, S.A., located in Leiria, Portugal). The recycling process starts by packing film waste collection (municipal waste) which is then pressed into bales that are transported to the plant. There, the bales are broken, and the following steps are conducted to produce LDPE pellets: sorting (hand picking), grading (size reduction, film is cut into flakes), washing, drying, and extrusion (flakes are melted and extruded into pellets). This study uses flakes (Figure 2), collected after the drying process, instead of plastic pellets, reducing thus the recycling costs and easing the plastic addition during the AC mixture blending (dry process).

Figure 2. Flakes of plastic film collected as urban waste.

The quantity of flakes to add was defined using Marshall test results and volumetric properties [54]. Fonseca, Almeida, Capitão, Bandeira, and Rodrigues [54] tested an AC mixture with the same grading considering 0%, 2%, 4%, 6%, and 8% of flakes of plastic film, by weight of bitumen. The mixture with 6% presented better results (higher stability and lower flow) and consequently that was the percentage considered.

2.1.3. Nanoclay

The nanoclay is a hydrophilic bentonite ($H_2Al_2O_6Si$) with beige colour, a molecular mass of 180.1 g/mol, a density of 2,400 kg/m^3, and pH in the range of 6.0 to 9.0. The nanometric dimension of the nanoclay is the thickness of its silicate layers of about 1 to 2 nm. In literature, nanoclay content range from 1% to 30% by weight of bitumen [43], being the mean value about 3.7%. This study considers 4.0% to be ideal, as in previous studies [42,50]. The blending procedure of nanoclay with the bitumen is detailed in [42]. Briefly, the procedure involved the use of a mechanical stirring effect to achieve an adequate dispersion of the nanoparticles into the bitumen.

2.2. Methods

For comparison purposes, three AC mixtures were produced: (1) a control one, without plastic and nanoclay; (2) a mixture containing only flakes of plastic film; and, (3) a mixture with both flakes of plastic film and nanoclay. The AC mixture design, in Portugal, is based on the Marshall method. However, as this study deals with a typical AC surface mixture whose binder content is well-established, the Marshall method was not specifically carried out once this was done before for the same materials. The binder content considered in this study was thus 5.0% by the weight of the mixture [54–56].

The moisture susceptibility of the compacted AC mixtures was evaluated using the indirect tensile strength ratio (ITSR) considering unaged and aged conditions (TEAGE procedure). Therefore, for each one of the three AC mixtures tested, 16 (2 ageing conditions × 2 ITSR conditions × 4 specimens) cylindrical specimens (63.5 mm height and 101.5 mm diameter) were produced and tested, which gives a total of 48 specimens. All the AC mixtures were produced in the laboratory at a target temperature of 165 °C accordingly the EN12697-35 [55], and the specimens were prepared by impact compactor by applying 75 blows on each side [56].

The bulk density of each specimen and the maximum density of each mixture were determined following the EN12697-6 [57] and the EN12697-5 [58], respectively. Figure 3 summarizes the tested AC mixtures.

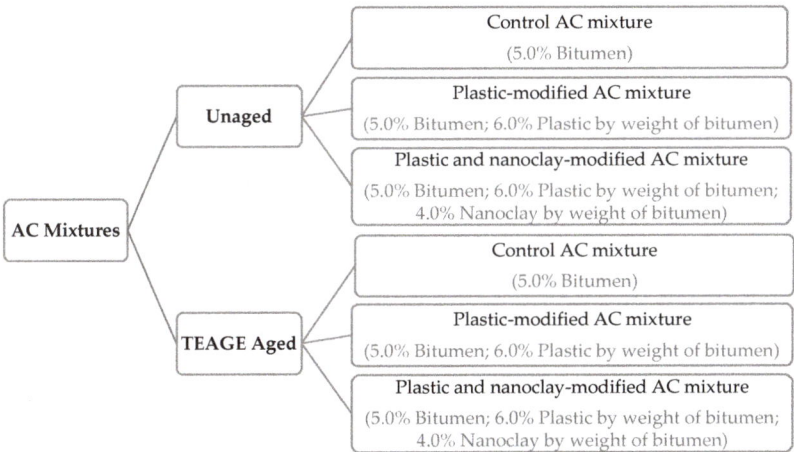

Figure 3. Tested asphalt concrete (AC) mixtures.

2.2.1. Ageing Conditioning

The TEAGE method was used to simulate in the laboratory the ageing of the asphalt mixtures [29]. Taking into account the historical climate data, TEAGE was tuned to simulate the effects of UV radiation and precipitation that the pavement in service will undergo in the Lisbon region (Portugal) during seven years. According to the Köppen–Geiger climate classification, Lisbon has a temperate dry hot-summer climate (classification Csa) with no freeze–thaw cycles. Thus, the UV radiation and the moisture damage will be important mechanism regarding the ageing of the asphalt pavement. Due to solar radiation, Lisbon receives an average annual energy of 5.7 GJ/m². Regarding precipitation, Lisbon has in average 40 days per year with rainfall higher than 5 mm. This amount of daily precipitation, 5 mm, is considered to be sufficient to cause a water flow in the pavement surface. The TEAGE prototype uses UV lamps to apply an equivalent level of energy and uses a combination of watering/drying cycles to simulate the effect of precipitation. The duration and details of the ageing conditioning are presented in Figure 4. From Figure 4, the effect of UV radiation on the colouration of the AC mixture specimens is visible.

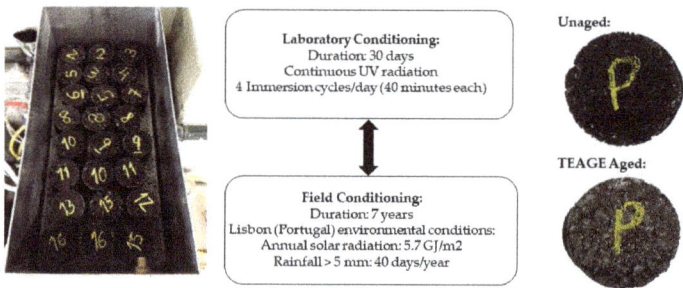

Figure 4. Tecnico accelerated ageing (TEAGE) simulation.

2.2.2. Moisture Sensitivity

Indirect tensile strength (ITS) tests were used to evaluate the moisture sensitivity by calculating the indirect tensile strength ratio ITSR as described in the EN 12697-12 [59]. The test temperature was 25 °C which is the recommended standard temperature in the EN 12697-12 [41]. ITSR is the ratio between the ITS of conditioned (wet) specimens, by immersion in water at 40 °C for approximately 72 h, and the ITS of unconditioned (dry) specimens. The ITS results are the average of four individual specimens per group. The wet set of specimens was previously subjected to vacuum with an absolute pressure of 6.7 kPa for a period of 30 min.

The ITS is defined as the maximum tensile stress calculated as a function of the peak load and the dimensions of the specimen [60], Equation (1):

$$ITS = \frac{2P}{\pi DH} \quad (1)$$

where ITS is the indirect tensile strength in GPa, P is the peak load in kN, D and H are the diameter and the height of the specimen in mm, respectively.

3. Results and Discussion

3.1. Volumetric Characterisation

The volumetric characterization is supported by the analysis of the air voids content of each specimen which was determined according to EN 12697-8 [61]. Figure 5 presents average bulk density values and air voids content (Vm) values for each AC mixture and set of specimens (dry/wet and unaged/aged). The vertical error bar represents the standard deviation of four specimens.

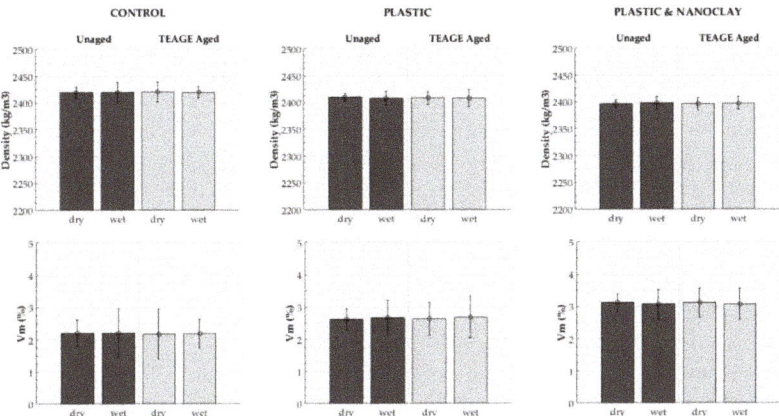

Figure 5. Bulk density and air voids content results for each AC mixture and set of specimens (dry/wet and unaged/aged).

From Figure 5, it is possible to observe the following:

- For each AC mixture, all sets have similar volumetric properties.
- The air voids contents of the tested AC mixtures are relatively low, which can be justified by the impact energy compaction used (75 blows) and by the fact that it is a dense-graded AC mixture.
- There was a decrease in the bulk density, and consequently an increase in air voids content, with the addition of plastic, as would be expected, since the plastic was used in replacement of bitumen, which is a denser material than the plastic.
- The specimens with plastic and nanoclay present higher air voids content. This can be partially explained by the increase of bitumen viscosity caused by the nanoclay.

3.2. Moisture Sensitivity and Ageing Effect

The ITSR of AC mixtures is an indicator of their resistance to moisture susceptibility. However, before presenting ITSR results, average ITS values for each case (AC mixture, dry vs wet, and unaged vs TEAGE aged) are presented in Table 1, depicted in Figure 6 and then discussed.

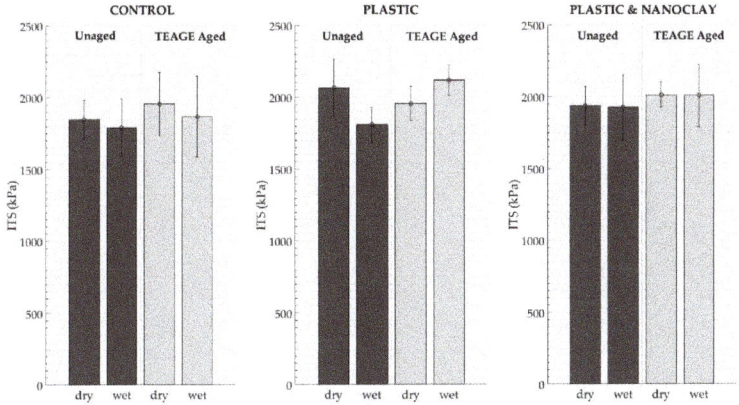

Figure 6. Indirect tensile strength (ITS) results.

Table 1. Indirect tensile strength (ITS) values.

ITS Values	Control				Plastic				Plastic & Nano			
	Unaged		TEAGE Aged		Unaged		TEAGE Aged		Unaged		TEAGE Aged	
	Dry	Wet	Dry	Wet	Dry	Wet	Dry	Wet	Dry	Wet	Dry	Wet
in kPa	1845	1790	1956	1867	2065	1808	1957	2118	1936	1926	2013	2007
in psi	268	260	284	271	300	262	284	307	281	279	292	291

From Figure 6, it is possible to observe the following:

- All the AC mixtures showed high ITS values even in wet specimens. It can be justified by the reduced air voids content of the samples that minimizes the moisture damage of the mixture [23], and by the bitumen grade used (low penetration grade) [21].
- Notwithstanding the slightly higher air voids content, both modified AC mixtures presented an increase in the conventional ITS (unaged and dry conditions). Compared with the ITS of the control AC mixture, the plastic modified AC mixture presented an increase of 12%, and the plastic and nanoclay-modified AC mixture presented an increase of 5%.
- In general, ITS values of wet specimens (where moisture damage is present) are lower than the dry ones (no moisture damage).
- Ageing tends to increase ITS values as bitumen became stiffener during the ageing process [29,50]. The exception, on average, was the dry set of specimens of the plastic modified-AC mixture, which is more a verification of the fact for this set of tests than a documented trend, once the variability of the samples could induce some "not normal" results as can be understood by the interval of results found in Figure 6.
- The outcome of the ageing effect in the plastic modified-AC mixture is not obvious. The TEAGE aged dry specimens presented lower values (5%) than the unaged ones, and the TEAGE aged wet specimens presented higher values (8%) than the TEAGE aged dry ones.
- Nanoclay addition improved the resistance of plastic modified-AC mixture to moisture damage. The ITS values were similar in dry and wet specimens. The ageing effect on the plastic and nanoclay-modified AC mixture was less pronounced than in the other specimens.

A two-way analysis of variance (ANOVA) with interaction on the obtained ITS values was conducted. The response variable was ITS (dry and wet) and the influence factors were AC mixture and ageing. The ANOVA results are shown in Table 2. From the "p-value" column, which presents the statistical significance level of the two-way ANOVA. It is observed that the p-values of the influence of ageing, AC mixture, and ageing-AC mixture interaction (AB) are higher than the threshold value of 0.05. Therefore, those factors do not have a significant effect on results. Only for wet specimens, the ageing effect was pronounced (p-value equal to 0.075).

Table 2. Two-way ANOVA with interaction results.

Source	Sum of Squares		DF	Mean Square		F-Value		p-Value	
	Dry	Wet		Dry	Wet	Dry	Wet	Dry	Wet
A-Ageing	4085.5	145,359.6	1	4085.5	145,359.6	0.169	3.585	0.686	0.075
B-AC Mixture	50,538.5	99,090.9	2	25,269.2	49,545.4	1.045	1.222	0.372	0.318
AB	55,275.6	71,102.0	2	27,637.8	35,551.0	1.143	0.877	0.341	0.433
Error	435,115.0	729,880.9	18	24,173.1	40,548.9				
Total	545,014.6	1,045,433.4	23						

After carrying out a one-way ANOVA to evaluate only the ageing effect (Table 3), it is observed that it was only significant for the plastic-modified mixture (p-value equal to 0.008).

Table 3. One-way ANOVA (*p*-values).

Control		Plastic		Plastic and Nanoclay	
Dry	Wet	Dry	Wet	Dry	Wet
0.4252	0.6727	0.3835	0.008	0.3765	0.6261

Therefore and as already mentioned, the effect of ageing in the plastic modified-AC mixture is not obvious. On the one hand, for wet specimens, the ITS has a statically significant increase, on the other, for dry specimens, the ITS has a non-statically significant decrease. The unaged dry specimens results are in line with a previous study [54] that evaluated the moisture susceptibility effect on AC mixtures containing flakes of plastic film collected as urban waste without ageing. Exposure to moisture reduced ITS values by 10% as in this research (12%). The effect of ageing on the plastic-modified materials are not well established yet. Al-Hadidy [41] studied the effect of two ageing levels (1: 100 °C for 48 h and 2: 100 °C for 96 h) on ITS results of AC mixtures containing PP polymers (wet process) and compared the results with the ones from a control mixture. ITS values increased with ageing. However, those variations depended on the ageing level. For the 96 h-ageing level, the increase was less pronounced for the AC mixtures containing PP polymers. In the Al-Hadidy study, only the temperature effect was evaluated and for a short period of time. In the TEAGE ageing procedure, the specimens (both the dry and the wet ones) are subjected to temperature, watering/drying cycles, and UV radiation during 30 days, which may have a particular role in deteriorating the material (especially plastic film) that thermal ageing may not reproduce accurately.

Concerning the isolated effect of nanoclay modification, López-Montero, Crucho, Picado-Santos, and Miró [50] assessed the effect of nanoclay on ageing and moisture damage of a gap-graded AC 14 mixture using the ITS test, and compared the results with the ones from a non-modified mixture. They found a significant increase in ITS values with nanoclay modification and with ageing in both dry and wet conditions. Similar results were reported in [52].

Figure 7 presents ITSR values that represent the ratio between the ITS of wet specimens and the ITS of dry specimens presented in Figure 6.

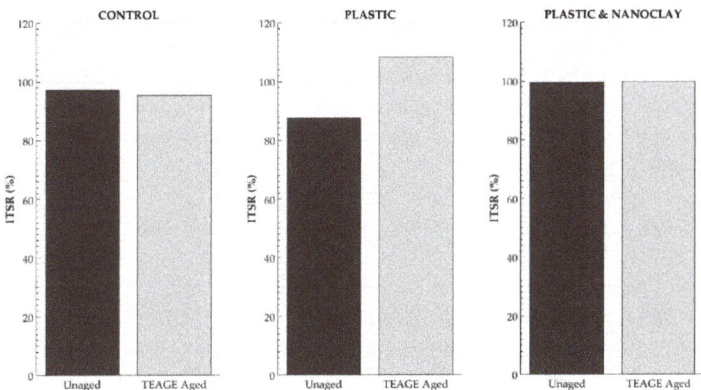

Figure 7. Indirect tensile strength ratio (ITSR) values.

All ITSR values are considered to be good, they are all above 85% whatever the AC mixture, indicating greater resistance to moisture damage. Once again, it can be explained by the air voids content values as well as by good affinity between the aggregates and the bitumen. The ageing considerably increased (21%) the moisture resistance of plastic modified-AC mixture. For the other two mixtures, the ITSR was near 100% and consequently the water did not deteriorate nor improve them and the ageing effect was minimal.

Diab, Enieb, and Singh [26] studied, among others, the influence of ageing (oven ageing during 16 h) on the ITSR of polymer-modified AC mixtures (six different polymeric products were tested). They found that polymeric products have different moisture susceptibility and that oven ageing can have different effects. In some modifications ITSR values increase while in others they decrease. These results suggest that the behaviour of plastic-modified AC mixtures might be complex and in some cases the expected trends could be not valid to reach bold conclusions.

In order to differently appreciate the effect of ageing on ITS values by separating dry and wet samples, an ageing index was calculated. This index is the quotient between the ITS of aged specimens and the ITS of unaged specimens [50]. Figure 8 shows the results.

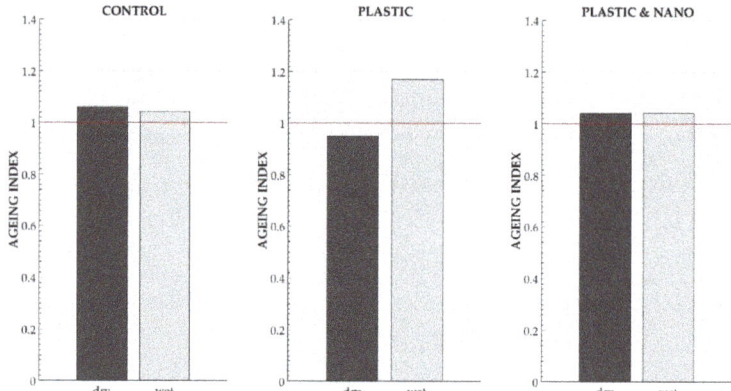

Figure 8. Values of ageing index for the tested mixtures and dry/wet specimens.

There was almost no difference for the control and plastic and nanoclay-modified AC mixtures. That was not the case for the plastic-modified mixture and as aforementioned this could be due to some variability of the samples, especially for the results of the dry set of samples, once the wet group showed the same trend than the others AC mixtures, namely having an ageing index greater than one. Some further testing framework will be needed, namely analysing more samples in the same context, different compositions with plastic, and different percentages of plastic content, but also appreciate the effect through the use of performance tests such the ones allowing comparison for fatigue and permanent deformation contexts, where the effect of ageing could be better separated.

4. Conclusions

The main objective of this study was to investigate the moisture susceptibility (by ITSR values) and the ageing effect (TEAGE procedure) of nanoclay-modified AC mixtures containing flakes of plastic film collected as urban waste. Three AC mixtures were produced: (1) a control one, without plastic and nanoclay; (2) a mixture containing only flakes of plastic film; and, (3) a mixture with both flakes of plastic film and nanoclay. In addition, aged and unaged specimens were considered.

After testing the AC mixtures and comparing the results, the following conclusions can be drawn:

- In general, the conditioning for moisture sensitivity caused a decrease in the ITS of the AC mixtures, and the ageing conditioning caused an increase in the ITS of the AC mixtures.
- The use of plastic to modify the AC mixture caused an increase of the ITS values in both ages (unaged and aged) and moisture conditioning (dry and wet), except in the case of the aged dry group where the ITS values were similar. However, the trends observed regarding moisture susceptibility and effect of ageing were not clear.
- The plastic and nanoclay-modified AC mixture presented higher ITS values, lower moisture susceptibility, and enhanced ageing resistance. Regarding moisture susceptibility, the mixture

presented similar ITS values for the dry and wet specimens, and consequently high ITSR values, indicating that the moisture conditioning had no effect on the properties of the mixture. Regarding ageing, if compared with the control mixture, the plastic and nanoclay-modified AC mixture presented slightly lower ageing index. The improvements obtained can be considered as relatively small, although, it worth mention that the plastic and nanoclay-modified AC mixture had an air void content higher than the control AC mixture, thus theoretically more vulnerable to moisture and ageing effect. Thus, the results obtained from the side of the modified AC mixture must be considered as conservative.

- The reduced air voids content of the AC mixtures tested certainly influenced the results. Lower compaction energy would have led to a higher air voids content and consequently, the moisture susceptibility, as well as the ageing effect, would be higher.

Therefore, the modification of AC mixtures with flakes of plastic and nanoclay hydrophilic bentonite might be a viable solution for the recycling of plastic film collected as urban waste. From the environmental point of view, it is an eco-friendly alternative to plastic waste disposal in landfills.

Regarding economic aspects, nanoclay modification is an expensive way to improve AC mixture performance. In fact, the construction cost can increase up to five times in relation to a conventional AC mixture [42]. Furthermore, in spite of plastic film and nanoclay together do not constitute today an industrial alternative to other types of modified mixtures in order to increase performance, this paper underlines that it is a trend with potential. Nanoclay plus plastic film could improve the performance and reduce the environmental burden, using an attractive mixture, even on the life-cycle cost point of view, especially if all the environmental costs are considered.

Although the reported results give a good indication of the potential of the modifications in evaluation, future research is recommended to see what happens in an AC mixture with a higher air voids content as well as to enhance the effect of ageing and moisture on plastic-modified AC mixtures.

Author Contributions: Conceptualization, A.A., J.C., C.A. and L.P.S.; methodology, A.A., J.C., C.A. and L.P.S.; validation, A.A., J.C., C.A. and L.P.S.; investigation, A.A., J.C. and C.A.; resources, A.A., J.C. and C.A.; writing—original draft preparation, A.A. and C.A.; writing—review and editing, A.A., J.C. and L.P.S.; supervision, L.P.S.; project administration, L.P.S.

Funding: This research received no external funding.

Conflicts of Interest: The authors declare no conflict of interest.

References

1. PlasticsEurope. *Plastics—The Facts 2018—An Analysis of European Plastics Production, Demand and Waste Data*; Association of Plastics Manufactures: Brussels, Belgium, 2018.
2. Statista. Global Plastic Production from 1950 to 2017. Available online: https://www.statista.com/statistics/282732/global-production-of-plastics-since-1950/ (accessed on 6 September 2019).
3. Mepex Consult AS. *Basic Facts Report on Design for Plastic Packaging Recyclability*; Mepex Consult AS: Asker, Norway, 2017.
4. Horodytska, O.; Valdés, F.J.; Fullana, A. Plastic flexible films waste management—A state of art review. *Waste Manag.* **2018**, *77*, 413–425. [CrossRef] [PubMed]
5. European Commission. *Plastic Waste: Ecological and Human Health Impacts*; Science Communication Unit, the University of the West of England (UWE), Ed.; European Commission's Directorate-General Environment: Bristol, UK, 2011; (In-depth Reports).
6. European Commission. *Report from the Commission to the European Parliament, the Council, the European Economic and Social Committee and the Committee of the Regions on the Implementation of EU Waste Legislation, Including the Early Warning Report for Member States at Risk of Missing the 2020 Preparation for Re-Use/Recycling Arget on Municipal Waste COM(2018) 656 Final*; European Commission: Brussels, Belgium, 2018.
7. European Commission. *Communication from the Commission to the European Parliament, the Council, the European Economic and Social Committee and the Committee of the Regions—A European Strategy for Plastics in a Circular Economy—COM/2018/028 Final*; European Commission: Brussels, Belgium, 2018.

8. Crippa, M.; De Wilde, B.; Koopmans, R.; Leyssens, J.; Muncke, J.; Ritschkoff, A.-C.; Van Doorsselaer, K.; Velis, C.; Wagner, M.A. *A Circular Economy for Plastics—Insights from Research and Innovation to Inform Policy and Funding Decisions*; Directorate-General for Research and Innovation European Commission: Brussels, Belgium, 2019.
9. Movilla-Quesada, D.; Raposeiras, A.C.; Silva-Klein, L.T.; Lastra-González, P.; Castro-Fresno, D. Use of plastic scrap in asphalt mixtures added by dry method as a partial substitute for bitumen. *Waste Manag.* **2019**, *87*, 751–760. [CrossRef] [PubMed]
10. Ziari, H.; Nasiri, E.; Amini, A.; Ferdosian, O. The effect of EAF dust and waste PVC on moisture sensitivity, rutting resistance, and fatigue performance of asphalt binders and mixtures. *Constr. Build. Mater.* **2019**, *203*, 188–200. [CrossRef]
11. Karmakar, S.; Majhi, D.; Roy, T.K.; Chanda, D. Moisture Damage Analysis of Bituminous Mix by Durability Index Utilizing Waste Plastic Cup. *J. Mater. Civ. Eng.* **2018**, *30*, 04018216. [CrossRef]
12. Martin-Alfonso, J.E.; Cuadri, A.A.; Torres, J.; Hidalgo, M.E.; Partal, P. Use of plastic wastes from greenhouse in asphalt mixes manufactured by dry process. *Road Mater. Pavement Des.* **2019**. [CrossRef]
13. Sarang, G.; Lekha, B.M.; Krishna, G.; Ravi Shankar, A.U. Comparison of Stone Matrix Asphalt mixtures with polymer-modified bitumen and shredded waste plastics. *Road Mater. Pavement Des.* **2016**, *17*, 933–945. [CrossRef]
14. Sarang, G. Replacement of stabilizers by recycling plastic in asphalt concrete. In *Use of Recycled Plastics in Eco-efficient Concrete*; Pacheco-Torgal, F., Khatib, J., Colangelo, F., Tuladhar, R., Eds.; Woodhead Publishing: Cambridge, UK, 2019; pp. 307–325.
15. Mishra, B.; Gupta, M.K. Use of plastic waste in bituminous mixes by wet and dry methods. *Inst. Civ. Eng. Munic. Eng.* **2018**, *0*, 1–11. [CrossRef]
16. Giri, J.P.; Panda, M.; Sahoo, U.C. Use of waste polyethylene for modification of bituminous paving mixes containing recycled concrete aggregates. *Road Mater. Pavement Des.* **2018**, 1–21. [CrossRef]
17. Vila-Cortavitarte, M.; Lastra-González, P.; Calzada-Pérez, M.Á.; Indacoechea-Vega, I. Analysis of the influence of using recycled polystyrene as a substitute for bitumen in the behaviour of asphalt concrete mixtures. *J. Clean Prod.* **2018**, *170*, 1279–1287. [CrossRef]
18. Leng, Z.; Sreeram, A.; Padhan, R.K.; Tan, Z. Value-added application of waste PET based additives in bituminous mixtures containing high percentage of reclaimed asphalt pavement (RAP). *J. Clean. Prod.* **2018**, *196*, 615–625. [CrossRef]
19. Jeong, K.-D.; Lee, S.-J.; Kim, K.W. Laboratory evaluation of flexible pavement materials containing waste polyethylene (WPE) film. *Constr. Build. Mater.* **2011**, *25*, 1890–1894. [CrossRef]
20. Nouali, M.; Derriche, Z.; Ghorbel, E.; Chuanqiang, L. Plastic bag waste modified bitumen a possible solution to the Algerian road pavements. *Road Mater. Pavement Des.* **2019**, 1–13. [CrossRef]
21. Zhang, J.; Apeagyei, A.K.; Airey, G.D.; Grenfell, J.R.A. Influence of aggregate mineralogical composition on water resistance of aggregate–bitumen adhesion. *Int. J. Adhes. Adhes.* **2015**, *62*, 45–54. [CrossRef]
22. Do, T.C.; Tran, V.P.; Le, V.P.; Lee, H.J.; Kim, W.J. Mechanical characteristics of tensile strength ratio method compared to other parameters used for moisture susceptibility evaluation of asphalt mixtures. *J. Traffic Transp. Eng.* **2019**. [CrossRef]
23. Terrel, R.L.; Al-Swailmi, S. *Water Sensitivity of Asphalt—Aggregate Mixes: Test Selection*; Strategic Highway Research Program, National Research Council: Washington, DC, USA, 1994.
24. Masad, E.; Castelblanco, A.; Birgisson, B. Effects of Air Void Size Distribution, Pore Pressure, and Bond Energy on Moisture Damage. *J. Test. Eval.* **2006**, *34*, 15–23.
25. Arambula, E.; Masad, E.; Martin, A.E. Influence of Air Void Distribution on the Moisture Susceptibility of Asphalt Mixes. *J. Mater. Civ. Eng.* **2007**, *19*, 655–664. [CrossRef]
26. Diab, A.; Enieb, M.; Singh, D. Influence of aging on properties of polymer-modified asphalt. *Constr. Build. Mater.* **2019**, *196*, 54–65. [CrossRef]
27. Airey, G.D.; Choi, Y.-K. State of the Art Report on Moisture Sensitivity Test Methods for Bituminous Pavement Materials. *Road Mater. Pavement Des.* **2002**, *3*, 355–372. [CrossRef]
28. Teh, S.Y.; Hamzah, M.O. Asphalt mixture workability and effects of long-term conditioning methods on moisture damage susceptibility and performance of warm mix asphalt. *Constr. Build. Mater.* **2019**, *207*, 316–328. [CrossRef]

29. Crucho, J.; Picado-Santos, L.; Neves, J.; Capitão, S.; Al-Qadi, I.L. Tecnico accelerated ageing (TEAGE)—A new laboratory approach for bituminous mixture ageing simulation. *Int. J. Pavement Eng.* **2018**, 1–13. [CrossRef]
30. Al-Hadidy, A.I. Effect of laboratory aging on moisture susceptibility and resilient modulus of asphalt concrete mixes containing PE and PP polymers. *Karbala Int. J. Mod. Sci.* **2018**, *4*, 377–381. [CrossRef]
31. AASHTO. *Standard Practice for Mixture Conditioning of Hot Mix Asphalt (HMA)—AASHTO Designation: R 30-02*; American Association of State Highway and Transportation Officials: Washington, DC, USA, 2015.
32. Smith Braden, T.; Howard Isaac, L. Comparing Laboratory Conditioning Protocols to Longer-Term Aging of Asphalt Mixtures in the Southeast United States. *J. Mater. Civ. Eng.* **2019**, *31*, 04018346. [CrossRef]
33. Zhang, H.; Zhu, C.; Yu, J.; Shi, C.; Zhang, D. Influence of surface modification on physical and ultraviolet aging resistance of bitumen containing inorganic nanoparticles. *Constr. Build. Mater.* **2015**, *98*, 735–740. [CrossRef]
34. Yi, M.W.; Wang, J.C.; Feng, X.D. Effect of Ultraviolet Light Aging on Fatigue Properties of Asphalt. *Key Eng. Mater.* **2014**, *599*, 125–129. [CrossRef]
35. Feng, Z.-G.; Bian, H.-J.; Li, X.-J.; Yu, J.-Y. FTIR analysis of UV aging on bitumen and its fractions. *Mater. Struct.* **2016**, *49*, 1381–1389. [CrossRef]
36. De Sá Araujo, M.d.F.A.; Lins, V.d.F.C.; Pasa, V.M.D.; Leite, L.F.M. Weathering aging of modified asphalt binders. *Fuel Process. Technol.* **2013**, *115*, 19–25. [CrossRef]
37. Islam, M.R.; Tarefder, R.A. Study of Asphalt Aging Through Beam Fatigue Test. *Transp. Res. Rec.* **2015**, *2505*, 115–120. [CrossRef]
38. Yin, F.; Arámbula-Mercado, E.; Epps Martin, A.; Newcomb, D.; Tran, N. Long-term ageing of asphalt mixtures. *Road Mater. Pavement Des.* **2017**, *18*, 2–27. [CrossRef]
39. Airey, G.D. State of the Art Report on Ageing Test Methods for Bituminous Pavement Materials. *Int. J. Pavement Eng.* **2003**, *4*, 165–176. [CrossRef]
40. Izadi, A.; Motamedi, M.; Alimi, R.; Nafar, M. Effect of aging conditions on the fatigue behavior of hot and warm mix asphalt. *Constr. Build. Mater.* **2018**, *188*, 119–129. [CrossRef]
41. Islam, M.R.; Hossain, M.I.; Tarefder, R.A. A study of asphalt aging using Indirect Tensile Strength test. *Constr. Build. Mater.* **2015**, *95*, 218–223. [CrossRef]
42. Crucho, J.M.L.; Neves, J.M.C.d.; Capitão, S.D.; Picado-Santos, L.G.d. Mechanical performance of asphalt concrete modified with nanoparticles: Nanosilica, zero-valent iron and nanoclay. *Constr. Build. Mater.* **2018**, *181*, 309–318. [CrossRef]
43. Martinho, F.C.G.; Farinha, J.P.S. An overview of the use of nanoclay modified bitumen in asphalt mixtures for enhanced flexible pavement performances. *Road Mater. Pavement Des.* **2019**, *20*, 671–701. [CrossRef]
44. Fang, C.; Yu, R.; Liu, S.; Li, Y. Nanomaterials Applied in Asphalt Modification: A Review. *J. Mater. Sci. Technol.* **2013**, *29*, 589–594. [CrossRef]
45. Li, R.; Xiao, F.; Amirkhanian, S.; You, Z.; Huang, J. Developments of nano materials and technologies on asphalt materials—A review. *Constr. Build. Mater.* **2017**, *143*, 633–648. [CrossRef]
46. Saltan, M.; Terzi, S.; Karahancer, S. Performance analysis of nano modified bitumen and hot mix asphalt. *Constr. Build. Mater.* **2018**, *173*, 228–237. [CrossRef]
47. Hamedi, G.H.; Moghadas Nejad, F.; Oveisi, K. Investigating the effects of using nanomaterials on moisture damage of HMA. *Road Mater. Pavement Des.* **2015**, *16*, 536–552. [CrossRef]
48. Hamedi, G.H.; Moghadas Nejad, F. Use of aggregate nanocoating to decrease moisture damage of hot mix asphalt. *Road Mater. Pavement Des.* **2016**, *17*, 32–51. [CrossRef]
49. Hamedi, G.H.; Nejad, F.M.; Oveisi, K. Estimating the moisture damage of asphalt mixture modified with nano zinc oxide. *Mater. Struct.* **2016**, *49*, 1165–1174. [CrossRef]
50. López-Montero, T.; Crucho, J.; Picado-Santos, L.; Miró, R. Effect of nanomaterials on ageing and moisture damage using the indirect tensile strength test. *Constr. Build. Mater.* **2018**, *168*, 31–40. [CrossRef]
51. Sezavar, R.; Shafabakhsh, G.; Mirabdolazimi, S.M. New model of moisture susceptibility of nano silica-modified asphalt concrete using GMDH algorithm. *Constr. Build. Mater.* **2019**, *211*, 528–538. [CrossRef]
52. Crucho, J.M.L.; Neves, J.M.C.d.; Capitão, S.D.; Picado-Santos, L.G.d. Evaluation of the durability of asphalt concrete modified with nanomaterials using the TEAGE aging method. *Constr. Build. Mater.* **2019**, *214*, 178–186. [CrossRef]
53. IP. *Paving Materials Specifications*; Infraestruturas de Portugal: Lisbon, Portugal, 2014. (In Portuguese)

54. Fonseca, M.; Almeida, A.; Capitão, S.; Bandeira, R.; Rodrigues, C. Avaliação do Uso de Plástico Recuperado de Resíduos Sólidos Urbanos Como Agente Modificador de Misturas Betuminosas. In Proceedings of the 9th Congresso Rodoviário Português (9th Portuguese Road Congress), Lisbon, Portugal, 28–30 May 2019. (In Portuguese)
55. CEN. *EN12697-35. Bituminous Mixtures—Test Methods for Hot Mix Asphalt—Part 35: Laboratory Mixing*; CEN: Brussels, Belgium, 2004.
56. CEN. *EN12697-30. Bituminous Mixtures—Test Methods for Hot Mix Asphalt—Part 30: Specimen Prepared by Impact Compactor*; CEN: Brussels, Belgium, 2012.
57. CEN. *EN12697-6. Bituminous Mixtures—Test Methods for Hot Mix Asphalt—Part 6: Determination of Bulk Density of Bituminous Specimens*; CEN: Brussels, Belgium, 2012.
58. CEN. *EN12697-5. Bituminous Mixtures—Test Methods for Hot Mix Asphalt—Part 5: Determination of the Maximum Density*; CEN: Brussels, Belgium, 2018.
59. CEN. *EN12697-12. Bituminous Mixtures—Test Methods for Hot Mix Asphalt—Part 12: Determination of the Water Sensitivity of Bituminous Specimens*; CEN: Brussels, Belgium, 2018.
60. CEN. *EN12697-23. Bituminous Mixtures—Test Methods for Hot Mix Asphalt—Part 23: Determination of the Indirect Tensile Strength of Bituminous Specimens*; CEN: Brussels, Belgium, 2017.
61. CEN. *EN12697-8. Bituminous Mixtures—Test Methods for Hot Mix Asphalt—Part 8: Determination of Void Characteristics of Bituminous Specimens*; CEN: Brussels, Belgium, 2012.

© 2019 by the authors. Licensee MDPI, Basel, Switzerland. This article is an open access article distributed under the terms and conditions of the Creative Commons Attribution (CC BY) license (http://creativecommons.org/licenses/by/4.0/).

Article

Evaluation of Rheological Behavior, Resistance to Permanent Deformation, and Resistance to Fatigue of Asphalt Mixtures Modified with Nanoclay and SBS Polymer

Gabriela Ceccon Carlesso *, Glicério Trichês, João Victor Staub de Melo, Matheus Felipe Marcon, Liseane Padilha Thives and Lídia Carolina da Luz

Department of Civil Engineering, Federal University of Santa Catarina, Florianópolis 88040-900, Brazil
* Correspondence: gabriela.carlesso@gmail.com; Tel.: +55-49-99133-0706

Received: 23 May 2019; Accepted: 27 June 2019; Published: 2 July 2019

Abstract: Fatigue cracking and rutting are among the main distresses identified in flexible pavements. To reduce these problems and other distresses, modified asphalt mixtures have been designed and studied. In this regard, this paper presents the results of a study on rheological behavior and resistance to permanent deformation and to fatigue of four different asphalt mixtures: (1) with conventional asphalt binder (CAP 50/70); (2) with binder modified by nanoclay (3% NC); (3) with binder modified by styrene–butadiene–styrene polymer (SBS 60/85); and (4) with binder modified by nanoclay and SBS (3% NC + 2% SBS). For this analysis, the mixtures were evaluated based on complex modulus, permanent deformation tests, and fatigue tests (4PB, in the four-point bending apparatus), with the subsequent application of numerical simulations. The results obtained show a better rheological behavior related to greater resistance to permanent deformation for the mixture 3% NC + 2% SBS, which could represent an alternative for roads where a high resistance to rutting is required. Otherwise, on fatigue tests, higher resistance was observed for the SBS 60/85 mixture, followed by the 3% NC + 2% SBS mixture. Nevertheless, based on the results of the numerical simulations and considering the possibility of cost reduction for the use of the 3% NC + 2% SBS mixture, it is concluded that this modified material has potential to provide improvements to the road sector around the world, especially in Brazil.

Keywords: modified asphalt mixtures; nanomaterials; polymers; rheological behavior; fatigue cracking; permanent deformation

1. Introduction

Rutting and fatigue cracking can be considered the main distresses of flexible pavements. Besides reducing the safety and comfort of roadway users, it results in a need for increased vehicle maintenance.

In the search for better performance of asphalt coatings in relation to these problems, besides a higher strength to reduce other defects, the use of modified asphalt mixtures is one option available. In this context, asphalt mixtures modified with polymers have been applied successfully to road engineering since the 1970s, when they were first used in Europe [1]. In recent years, with the advent of nanotechnology, modification with the use of nanomaterials has also gained the attention of the scientific community. Studies carried out with asphalt nanocomposites have demonstrated the good performance and potential of these materials in the paving sector [2–12]. More recently, the behavior of asphalt binders and mixtures modified with polymers and nanomaterials have been investigated. The results obtained in this line of research have also been positive [13–21]. However, most studies have been limited to the binders, and there is a need for more investigations that consider the mixtures, taking into account the interaction between the binder, the granulometry, and the aggregate characteristics. This is important and should be considered because the total composition of the asphalt mixture will be in contact with traffic and weather during the pavement's lifespan, although the binder characteristics have a prominent role.

Thus, considering the high performance of mixtures modified with the addition of polymers and the potential for the application of nanomaterials to road engineering, the aim of this study was to carry out a comparative analysis of the rheological behavior and resistance to permanent deformation and to fatigue of four different asphalt mixtures: (1) a reference mixture, produced with a conventional asphalt binder (CAP 50/70: Petroleum Asphalt Cement with a penetration range between 5.0 and 7.0 millimeters) [22]; (2) a mixture with binder modified by nanoclay (3% NC), produced in laboratory [22]; (3) a mixture with binder modified by the polymer styrene–butadiene–styrene (SBS 60/85, with a minimum softening point of 60 °C and a minimum elastic recovery of 85%), produced industrially [23]; and (4) a mixture with binder modified by nanoclay and SBS (3% NC + 2% SBS), also produced in the laboratory [24]. This analysis was carried out through complex modulus and fatigue tests (4 PB, in the four-point bending apparatus) and permanent deformation tests (in the LCPC (Central Laboratory of Bridges and Roads/Laboratoire Central des Ponts et Chaussées) traffic simulator). Besides evaluating the behavior of the complex modulus and the phase angle of the mixtures with variations in the test load frequency and temperature, the rheological study allowed the prediction of the resistance of asphalt mixtures, especially with regard to rutting.

2. Materials and Methods

2.1. Materials

2.1.1. Aggregates and Granulometric Composition

The aggregates used in the production of the asphalt mixtures are of basaltic origin. The characterization of these materials is provided in Table 1. It can be observed that the properties listed are in conformity with the criteria established by the Superpave methodology (whenever applicable), verifying the suitability of the aggregates for the formulation of the mixtures.

Table 1. Characterization of aggregates [22].

Property	Standard	Result	Superpave Criterion
Bulk density of coarse aggregate	ASTM C 127 [25]	2.953 g/cm^3	n/a
Apparent density of coarse aggregate	ASTM C 127 [25]	2.880 g/cm^3	n/a
Absorption of coarse aggregate	ASTM C 127 [25]	0.8%	n/a
Bulk density of fine aggregate	DNER-ME 084 [26]	2.974 g/cm^3	n/a
Bulk density of powdery material	DNER-ME 085 [27]	2.804 g/cm^3	n/a
Angularity of coarse aggregate	ASTM D 5821 [28]	100%/100%	100%/100% min. [1]
Angularity of fine aggregate	ASTM C 1252 [29]	49.2%	45% min.
Flat and elongated particles	ABNT NBR 6954 [30]	9.6%	10% max.
Clay content (Sand equivalent)	AASHTO T 176 [31]	61.2%	50% min.
Hardness (Los Angeles abrasion)	ASTM C 131 [32]	11.6%	35–45% max.
Soundness	ASTM C 88 [33]	2.1%	10–20% max.
Deleterious materials	AASHTO T 112 [34]	0%	0.2–10% max.

Note: [1] e.g.,: 85%/80% means that 85% of coarse aggregate has one or more fractured faces and 80% has two or more fractured faces.

Table 2 presents the characterization of the hydrated lime used in the study, which corresponds to type CH-1 dolomitic.

Table 2. Characterization of the hydrated lime [22].

Property	Result
Loss on ignition	18.6%
Insoluble residue	1.9%
Carbon dioxide (CO_2)	2.5%
Calcium oxide (CaO)	45.1%
Magnesium oxide (MgO)	33.5%
Total non-volatile oxides (CaO + MgO)	96.5%
Total non-hydrated oxides	27.6%
Non-hydrated CaO	0.0%
Calcium (Ca)	32.2%
Magnesium (Mg)	20.2%
Bulk Density	3.00 g/cm^3

The formulation of the asphalt mixtures, based on the granulometric curve shown in Figure 1, was comprised of 43% gravel, 15.5% of crushed gravel, 40% of grit, and 1.5% of lime. This composition was established by the Leopoldo Américo Miguez de Mello Research and Development Center (CENPES/Petrobras), which aimed at obtaining mixtures with a high resistance to permanent deformation. The granulometric curve, aggregates, and hydrated lime were chosen because they were also applied in an experimental monitored road stretch still under evaluation by the authors.

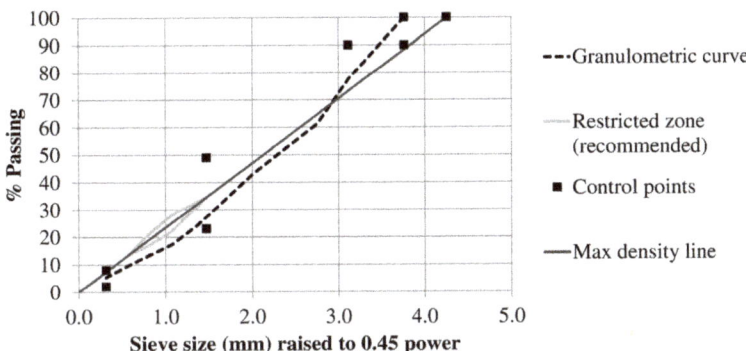

Figure 1. Granulometric curve for aggregate composition.

2.1.2. Conventional Asphalt Binder

The conventional asphalt binder used in this study was a CAP 50/70 (Petroleum Asphalt Cement with a penetration range between 5.0 and 7.0 millimeters), with PG 58–22 (Performance Grade 58-22). This binder was used in the production of the reference asphalt mixture and as a matrix for the modification of 3% NC and 3% NC + 2% SBS binders. The asphalt binder characterization is shown in Table 3.

2.1.3. Asphalt Binder Modified with SBS Polymer

The binder modified with the SBS polymer was industrially produced and supplied by Greca Asfaltos S.A. The characterization of this material is provided in Table 4.

Table 3. Characterization of the conventional asphalt binder (CAP 50/70).

Property	Unit	Standard	Result
Penetration	0.1 mm	ASTM D 5 [35]	57
Softening point	°C	ASTM D 36 [36]	47.9
Thermal susceptibility index	-	-	−1.44
Brookfield viscosity			
at 135 °C (*spindle* 21, 20 rpm)	cP	ASTM D 4402 [37]	290
at 150 °C (*spindle* 21, 50 rpm)			150
at 175 °C (*spindle* 21, 100 rpm)			60

Table 4. Characterization of the asphalt binder modified by styrene–butadiene–styrene (SBS) polymer (SBS 60/85).

Property	Unit	Standard	Result
Penetration	0.1 mm	ASTM D 5 [35]	50
Softening point	°C	ASTM D 36 [36]	73.0
Elastic recovery	%	ABNT NBR 15086 [38]	90
Apparent viscosity			
at 135 °C (*spindle* 21, 20 rpm)	cP	ASTM D 4402 [37]	1910
at 150 °C (*spindle* 21, 50 rpm)			640
at 175 °C (*spindle* 21, 100 rpm)			290

2.1.4. Modifiers

The organophillic nanoclay used in the modification of CAP 50/70 is known commercially as Dellite 67G. It has a particle size (dry) of 7–9 μm, a particle size after dispersion of 1 × 500 nm, and density of 1.7 g/cm^3. It has the following chemical composition: carbon (45.50%), silica (33.42%), aluminum (16.08%), iron (3.60%), chloride (0.80%), titanium (0.31%), potassium (0.27%), and strontium (0.02%). According to the results of thermogravimetry tests, this nanomaterial is thermally stable at temperatures below 262.4 °C [22].

The polymer SBS used as a modifier was Kraton D1101. It has a linear structure, with a polystyrene content of between 30% and 32%, and it was supplied in granules.

2.2. Methods

This study was carried out in seven stages.

Firstly, in Stage 1, the modification of the conventional binder and the characterization of the modified binders were carried out. In this stage, the binders 3% NC (modified by 3% nanoclay in relation to the weight of CAP 50/70) and 3% NC + 2% SBS (modified by 3% of nanoclay and 2% of SBS polymer) were obtained. The content of nanoclay was established as 3% based on the results of a previous optimization study developed by Melo [22], who evaluated the permanent deformation resistance of asphalt mixtures produced with binders modified by 1%, 2%, and 3% of nanoclay and verified a better performance for the 3% NC mixture. This nanoclay content was also identified in a recent study [39] as the most frequent value mentioned in the related literature. The addition of 2% SBS together with 3% NC was aimed at obtaining an asphalt material with elastic recovery. Low polymer content was adopted based on results reported by Pamplona et al. [20], who studied the modification with 2.5% of nanoclay and 2.5% of SBS, and also for economic reasons. It means the authors would like to produce a binder with reduced polymer content due to the cost of this material. However, another optimization of the modifier amount, including the SBS content, is recommended for future studies.

The inclusion of nanoclay and the polymer SBS in the base binder CAP 50/70 was carried out in a laboratory high shear mixer (Silverson, model L5M-A), and the modification procedures were defined based on studies reported in the literature [5,11,12,15–20]. The modification to produce the 3% NC binder was carried out at 150 °C, with a shearing speed of 5000 rpm and compatibilization period of 100 min. In this regard, the previously mentioned literature review study [39] also identified the following values as more frequent in similar studies: temperature of 160 °C, mixing speed of 4000 rpm, and mixing process duration of 120 min. This indicates this work is consistent with the literature, since the adopted and mentioned values are close. In the case of the 3% NA + 2% SBS binder, a modification temperature of 180 °C and a mixing period of 180 min were adopted, maintaining the shearing speed of 5000 rpm. These differences in time and temperature of modification were established in order to allow the complete dispersion of the modifiers. However, it should be noted that this can also do some influence on the binders and mixtures behavior. It is also interesting to mention that, as the binders were produced using a laboratory high shear mixer, for the reproduction of these materials in large quantities, with the same characteristics and aiming at the adequate dispersion of the modifiers, it would be necessary to use of an industrial high-shear mixer or another solution that provides the same results in terms of modifiers dispersion.

After production, the modified binders were then characterized according to the following properties: penetration (ASTM D 5 [35]), softening point ASTM D 36 [36]), elastic recovery (ABNT NBR 15086 [38]), phase separation (ABTN NBR 15166 [40]), and apparent viscosity (ASTM D 4402 [37]).

Stage 2 consisted of establishing the design binder contents of the asphalt mixtures, beginning the part of this study that aims to evaluate the influence of the modifiers when interacting in the asphalt mixtures. For this, the Superpave mix design method was applied, according to the standards AASHTO M 323 [41] and AASHTO R 35 [42], with the use of a gyratory compactor. Considering a maximum nominal size of the aggregate of 19 mm and a heavy volume of traffic, the binder contents of the design correspond to the following dosage criteria that were met simultaneously: (1) percentage of void volume in Ninitial (9 spins, Vv@Ninitial) > 11.0%; (2) percentage of void volume in Ndesign (125 spins, Vv@Ndesign) = 4.0%; (3) voids in mineral aggregate (VMA) ≥ 13.0%; (4) voids filled with asphalt (VFA) between 65% and 75%; and (5) dust to effective binder ratio between 0.8 and 1.6. It is important to note that, during the design mix and the subsequent steps, the asphalt mixtures were produced and compacted at temperatures of which the binder viscosities corresponded respectively to 0.17 Pa.s and 0.28 Pa.s.

In Stage 3, after the design binder content definition, for each mixture, three plates were molded with one having dimensions of 60 × 40 × 9 cm (for sawing and obtaining prismatic specimens for the complex modulus and fatigue tests) and two plates having dimensions of 50 × 18 × 5 cm (for the permanent deformation tests). This molding was carried out at the LCPC compacting table, following the recommendations of the French standard AFNOR NF P 98–250-2 [43] for a heavy traffic highway. Plates with dimensions 60 × 40 × 9 cm were then sawn, and five specimens with dimensions close to 6.3 × 5.0 × 40.0 cm were obtained from each plate.

In Stage 4, the rheological behavior characterization of the different mixtures was evaluated. This characterization was based on complex module tests, carried out in a four-point bending test machine, following the recommendations of the European standard EN 12697-26 [44]. For each asphalt mixture, two specimens produced in the previous stage were tested. These tests were carried out under alternating bending with the application of sinusoidal loading keeping the deformation amplitude at 50 µε. Temperatures of 0 °C, 5 °C, 10 °C, 15 °C, 20 °C, 25 °C, and 30 °C with load frequencies of 0.1 Hz, 0.2 Hz, 0.5 Hz, 1 Hz, 2 Hz, 5 Hz, 10 Hz, and 20 Hz, were evaluated. For the interpretation of the results, master curves and black spaces were analyzed. The master curves were built in the reference temperature of 20 °C, based on the TTS principle (time–temperature superposition) and by applying the Williams–Landel–Ferry equation, of which the constants were calculated using the Viscoanalyse software (developed by the LCPC).

In Stage 5, the resistance to permanent deformation of the four different mixtures was evaluated. In this stage, tests were performed in the French traffic simulator following the standard EN 12697-22+A1 [45]. These tests were carried out at 60 °C, with the application of a single axle with single wheel load, intensity of 5 kN, tire inflation pressure of 0.6 MPa, and frequency of 1 Hz.

The development of rutting recesses was measured after the following numbers of cycles: 100, 300, 1000, 3000, 10,000, and 30,000. The results of these tests after 30,000 cycles were compared to the limits established by the French [46] and European [47] guidelines.

Stage 6 consisted in evaluating the fatigue resistance of the asphalt mixtures. As mentioned for the rheological characterization, fatigue tests were also carried out in four-point bending test machine but followed the recommendations of the European standard EN 12697-24 [48]. In this stage of the study, about fifteen specimens were tested for each mixture under controlled deformation at deformation levels between 80 and 375 $\mu\varepsilon$ (μm/m). These tests were performed considering a frequency of 10 Hz (simulating the vehicles speed, in practice, of 72 km/h [49]) and at the temperatures between 15 °C and 20 °C (defined according to the average temperature of the region of this study). As a rupture criterion, it adopted a reduction of 50% of the initial complex modulus. At the end of the tests, it was possible to obtain the fatigue models of the mixtures, as presented in Equation (1):

$$N = k(\mu\varepsilon)^{-n}, \tag{1}$$

where N = number of cycles (loading applications) until asphalt concrete reaches 50% of initial stiffness; $\mu\varepsilon$ = maximum tensile strain applied on the material; and k, n = constants mainly dependent on stiffness and asphalt content of the mixture.

After that, in Stage 7, numerical simulations were carried out, considering the structures presented in Figure 2 (S1, S2, S3, and S4 with the same subgrades, subbases, and bases but different asphalt layers). These structures were chosen for analysis and comparisons because they represent variations of the structure S1, which was constructed in a roadway in the region where the authors developed this work and where the asphalt surface presents early distresses. This road stretch with the structure S1 is being monitored by the authors in other projects. The simulations aimed at estimating the tensile strain suffered by the lower fibers of the asphalt layers during the passage of vehicles (at 170 mm depth, according Figure 2). Thus, in the simulations, it was considered the load configuration shown in Figure 3, at the speed of 20 m/s and the fatigue tests' temperature. Applying this stage results to the fatigue models obtained in Stage 6, it was possible to estimate the lifespan in terms of asphalt concrete fatigue fracture of each structure analyzed. This procedure represents an initial estimation for comparisons, and lab to site shift factors can be applied in future researches. It also should be noted that the viscoelastic behavior of the asphalt mixtures was considered in the simulations by using the Huet-Sayegh rheological parameters, obtained from the results of Step 5.

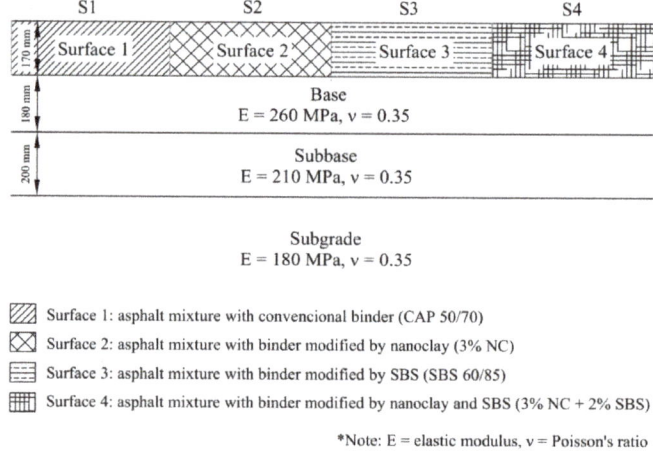

Figure 2. Pavement structures evaluated in the numerical simulations.

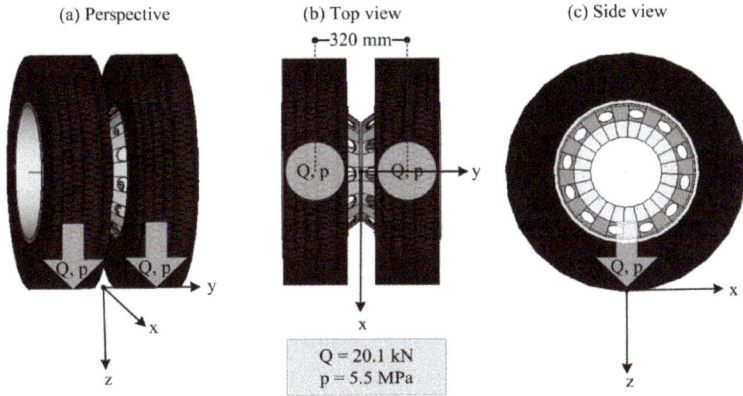

Figure 3. Load configuration considered for the numerical simulations.

3. Results and Discussion

3.1. Characterization of Modified Binders

Results of the laboratory modified asphalt binders' characterization can be observed in Table 5.

Table 5. Characterization of asphalt binders 3% NC and 3% NC + 2% SBS.

Propriety	Unit	Standard	Asphalt Binder	
			3% NC	3% NC + 2% SBS
Penetration	0.1 mm	ASTM D 5 [35]	55	36
Softening point	°C	ASTM D 36 [36]	50.2	56.9
Elastic recovery	%	ABNT NBR 15086 [38]	6	49
Phase separation (24 h/48 h)	°C	ABNT NBR 15166 [40]	1.0/-	0.5/0.8
Apparent viscosity	cP	ASTM D 4402 [37]		
at 135 °C (*spindle* 21, 20 rpm)			410	760
at 150 °C (*spindle* 21, 50 rpm)			210	370
at 175 °C (*spindle* 21, 100 rpm)			90	160

In comparing the results in Table 5 with those provided for the empirical characterization of CAP 50/70 (Table 3) and SBS 60/85 (Table 4), it can be observed that, as expected and also demonstrated by other authors in similar studies [13–21], the binders modification caused a penetration decrease and a softening point increase. Notable among these results are the relatively low penetration obtained for the binder 3% NC + 2% SBS and the relatively high softening point for the binder SBS 60/85. This reflects gains in the stiffness in the first case in contrast with gains related to the sensitivity to high temperatures in the second case. These characteristics indicate that the asphalt mixtures formulated will be very resistant to permanent deformation. Thus, according to the results obtained in the empirical characterization of the binders, better performance is expected in relation to the permanent deformation for the mixtures SBS 60/85 and 3% NC + 2% SBS. However, it is important to highlight that the behavior prediction of asphalt mixtures based on the modified binders' empiric characterization shows high limitations.

In relation to the elastic recovery property, the obtainment of a higher value is observed for the binder SBS 60/85. This result is due to the relatively high content of the elastomeric polymer added to the material (approximately 4%). On the other hand, the low elastic recovery of the binder 3% NC is related to the fact that, in the modification, the nanoclay does not function as an elastomeric product.

With regard to storage stability, the tests carried out with binders modified by nanoclay (3% NC and 3% NC + 2% SBS) revealed that they do not show significant phase separation. By way of comparison, it can be noted that the results are below the maximum limit (5 °C) established by the Brazilian Specification for Asphalt–Polymer [50].

In the same way, the results obtained for the modified binders' viscosity also met the limits established by the Brazilian Specification for Asphalt–Polymer [50] despite the considerable increases in comparison to CAP 50/70 (conventional).

3.2. Definition of Binder Contents

Table 6 shows the design binder contents obtained for the mixtures.

Table 6. Design binder contents of different asphalt mixtures.

Asphalt Binder	Design Binder Content (%)
CAP 50/70	4.4
3% NC	4.1
SBS 60/85	4.5
3% NC + 2% SBS	4.3

As can be observed in Table 6, the nanoclay addition to the conventional binder can lead to a reduction in the design binder content. In contrast, with the polymer SBS addition, higher binder content is required. Thus, the intermediate result obtained for the 3% NC + 2% SBS mixture in the design mix study demonstrates the combination of the nanoclay positive effect, in terms of mixture workability, with the negative effect of the presence of the polymer.

Also concerning the different binder contents obtained for the mixtures it should be noted that their rheological behavior, resistance to permanent deformation and resistance to fatigue (whose results will be presented later) can be influenced because the mixtures were not produced with exactly the same binder content (an exception to this comment are the mixtures CAP 50/70 and 3% NC, which were both produced with the same binder content = 4.4%). However, it's important to highlight the previously mentioned variation in the design binder contents is meeting the limit values allowed in the field (± 0.3%), based on Brazilian current construction specifications.

3.3. Plate Molding

Table 7 reports the results of void volume percentages checked for the specimens used in the complex modulus and fatigue tests. Table 8 also reports the results of void volume percentages but, in this case, obtained for the plates used in the permanent deformation tests.

Table 7. Volumetric characterization of specimens used in complex modulus and fatigue tests.

Asphalt Mixture	Void Volume (%)	
	Mean	Standard Deviation
CAP 50/70	4.53	0.87
3% NC	3.85	0.38
SBS 60/85	4.62	0.39
3% NC + 2% SBS	3.83	0.54

It should be noted that the plates molded to obtain specimens (modulus and fatigue tests) and those molded for permanent deformation tests were produced by aiming at 4% void volumes, as defined in the design mix study. However, during the compaction procedure, carried out at the LCPC compacting table (compaction procedure different from that used in the mixtures design step), it was found to be difficult to control the final thickness of the plates, hindering the precise obtainment of the void volumes equal to 4%. Even so, considering that the specimens and plates met the compaction degrees admitted in the field in Brazil (between 97% and 101%), they were considered suitable for use in the tests.

Table 8. Volumetric characterization of plates used in permanent deformation tests.

Asphalt Mixture	Plate	Void Volume (%)
CAP 50/70	P1	6.13
	P2	4.50
3% NC	P1	6.55
	P2	6.68
SBS 60/85	P1	5.87
	P2	5.57
3% NC + 2% SBS	P1	5.37
	P2	5.69

3.4. Rheological Characterization

Figure 4 shows the master curves obtained for the asphalt mixtures evaluation. In general, the positioning of the curves indicates higher stiffness for the modified asphalt mixtures. An exception, however, is the mixture SBS 60/85 which, at high load frequencies (equivalent to low temperatures), shows a very similar rheological behavior to the conventional mixture. In comparison with the other studied mixtures, the mixture 3% NC + 2% SBS showed notably high values for the complex modulus. This characteristic can be considered as an indication that the mixture will have greater resistance to permanent deformation but can be hampered in terms of fatigue resistance.

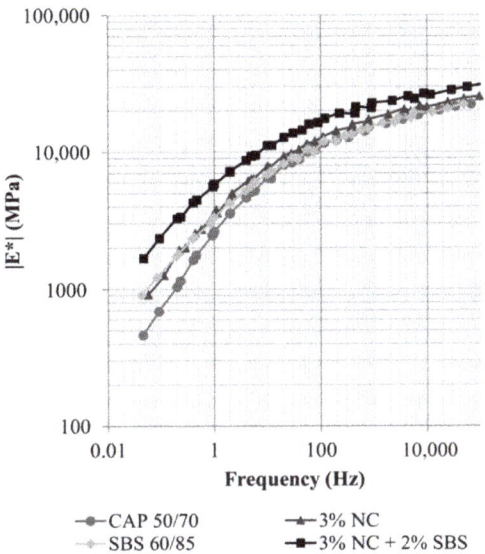

Figure 4. Master curves (reference temperature = 20 °C).

In addition, an analysis of the master curves also reveals the lower frequency susceptibility of the modified mixtures based on the slopes of the curves. In the field, this would mean that the stiffness of these mixtures should be less sensitive to variations in the traffic speed, suggesting a high potential for their use along segments with a steep slope and slow traffic. In this regard, it is possible to establish a behavior hierarchy, where the mixture modified with both modifiers is at the top, followed by SBS 60/85, 3% NC, and the conventional mixture.

It also can be noted in Figure 4 that the gains related to the frequency susceptibility of the modified mixtures are more significant at lower load frequencies (equivalent to higher temperatures). This may be considered a positive aspect in relation to the permanent deformation phenomenon in regions of tropical climate and on highways submitted to slow loads or in mountainous regions.

Figure 5 shows the black spaces for the different asphalt mixtures.

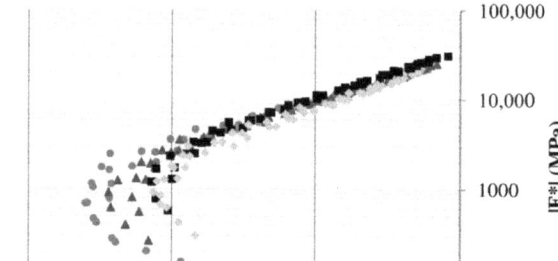
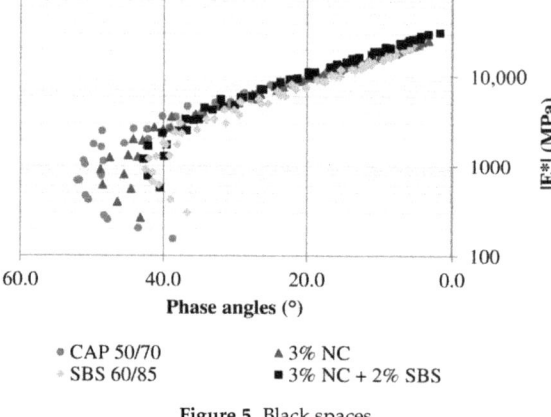

Figure 5. Black spaces.

It can be noted in Figure 5 that the modified mixtures present an aspect of graphic shortening (in relation to the phase angle) when compared with the conventional mixture. This shortening illustrates the obtainment of smaller phase angles and reflects the more elastic behavior of these mixtures. Of these, it can be observed that the 3% NC + 2% SBS mixture has a higher concentration of low phase angles, followed by the SBS 60/85 mixture. The elastic behavior of them, both modified with SBS, reflects the influence of the presence of an elastomeric polymer in the binders. Considered in isolation, this would suggest a better performance of these mixtures in relation to their resistance to permanent deformation and to fatigue cracking.

3.5. Resistance to Permanent Deformation

In Figures 6 and 7, the curves obtained in the permanent deformation tests and the results for the rutting depth (%) of mixtures after 30,000 cycles are provided, respectively.

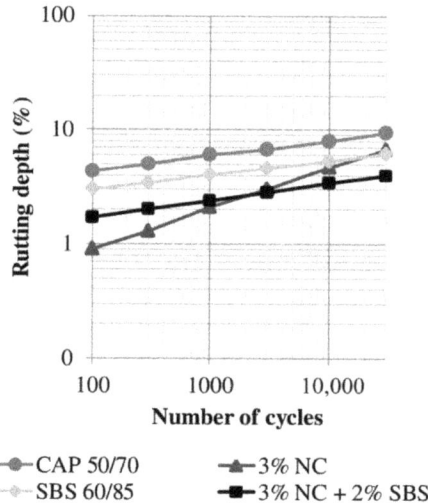

Figure 6. Curves obtained in permanent deformation tests.

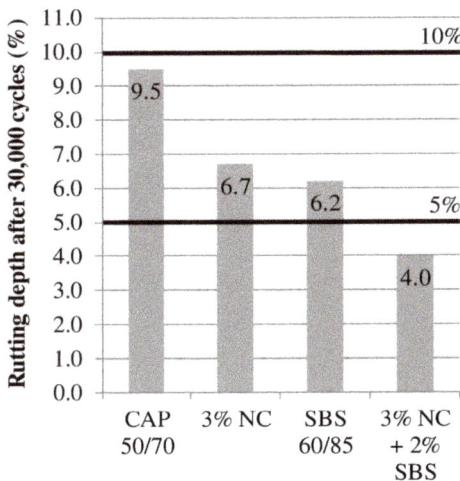

Figure 7. Results for rutting depth (%) after 30,000 cycles.

According to the results shown in Figures 6 and 7, the resistance to permanent deformation, considering 30,000 load cycles was highest for the 3% NC + 2% SBS mixture, followed by the SBS 60/85, 3% NC, and conventional mixtures. In proportional terms, in practice, it is estimated that the replacement of the conventional mixture containing the two modifiers would reduce the rutting depth in the field by up to 58%. Considering the replacement of the reference mixture with the SBS 60/85 and 3% NC mixtures, the corresponding reductions could reach up to 35% and 29%, respectively.

The best behavior in terms of permanent deformation observed for 3% NC + 2% SBS mixture is in agreement with the performance prediction carried out in the empirical characterization of the binders (lower penetration) and the rheological study of the mixtures (higher stiffness and lower phase angles). In the same way, the results for the phase angles were also effective in the prediction of the performance hierarchy of the mixtures in relation to resistance to permanent deformation.

About the rate of deformation increase, which can be measured by the WTS (wheel-tracking slope) parameter, Figure 6 shows similar behaviors between the mixtures CAP 50/70, SBS 60/85, and 3% NC + 2% SBS. On the other hand, it is observed that the 3% NC mix exhibits a lower initial deformation with a higher deformation increase rate. In this sense, considering about 3000 load cycles, for example, the deformation level was the same for the mixtures 3% NC and 3% NC + 2% SBS; however, due to a higher deformation increase of the mixture modified only by nanoclay, the scenario becomes different at the end of the test. High rates of deformation increase are considered a negative characteristic for asphalt mixtures performance, and according to the results obtained in this work, simultaneous action of SBS can improve this behavior of the 3% NC mixture.

Regarding the maximum limits for the rutting depth established by the French [46] and the European [47] specifications, all of the asphalt mixtures under study met the 10% criterion for 30,000 load cycles. However, only the 3% NC + 2% SBS mixture satisfied the more restrictive criterion, corresponding to 5%.

3.6. Resistance to Fatigue Cracking

The results of the fatigue tests are shown in Figure 8. Below the graph, the representative fatigue models of the mixtures are presented alongside the respective coefficients of determination (R^2) and the specific deformations for 10^6 cycles (ε_6). The models relate the tensile strain ($\mu\varepsilon$) applied on the specimens to the number of cycles (N, loading applications) until asphalt concrete reaches 50% of initial stiffness.

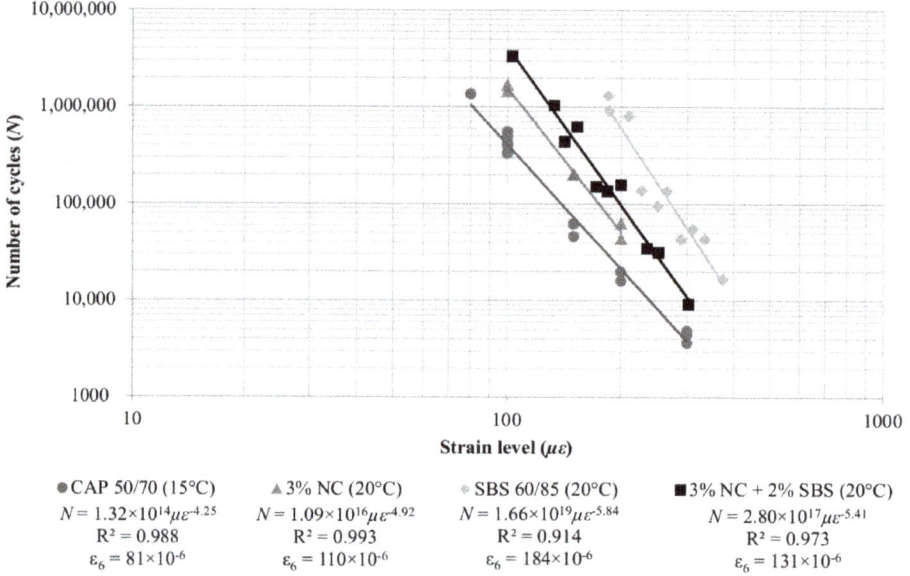

Figure 8. Curves obtained in fatigue tests.

For the conditions adopted in the tests carried out in this research, from the curves shown in Figure 8, it can be observed the higher fatigue strength of the SBS 60/85 mixture, followed by the 3% NC + 2% SBS, 3% NC and conventional mixtures.

Concerning the fatigue behavior prediction based on the rheological study, it can be noted that it was partially accomplished. In this sense, the superior behavior of the SBS 60/85 and 3% NC + 2% SBS mixtures was expected, due to the greater predominance of the elastic behavior (lower phase angles). However, according to the respective results, it was expected that the best performance would be of the 3% NC + 2% SBS mixture, in comparison to the SBS 60/85, which did not occur. Probably, in terms of fatigue, the mixture modified by nanoclay and SBS was impaired due to the considerable increase in the stiffness of the material, which was verified in the empirical characterization of the binder (low penetration) and also in the rheological study of the mixtures.

3.7. Results of Numerical Simulations

Table 9 presents the numerical simulation results applied to the fatigue models.

Table 9. Numerical simulations results applied to the fatigue models.

Structure/Asphalt Surface	Strain ($\mu\varepsilon$)	Fatigue Model	Number of Cycles (N)
S1/Surface with CAP 50/70	65	$N = 1.32 \times 10^{14} \, \mu\varepsilon^{-4.25}$	2.60×10^6
S2/Surface with 3% NC	75	$N = 1.09 \times 10^{16} \, \mu\varepsilon^{-4.92}$	6.49×10^6
S3/Surface with SBS 60/85	80	$N = 1.66 \times 10^{19} \, \mu\varepsilon^{-5.84}$	1.28×10^8
S4/Surface with 3% NC + 2% SBS	61	$N = 2.80 \times 10^{17} \, \mu\varepsilon^{-5.41}$	6.15×10^7

From the results presented in Table 9, the following hierarchy can be established for the asphalt mixtures, in terms of fatigue lifespan, starting by the longest one: structure S3 (surface with SBS 60/85 mixture), structure S4 (surface with 3% NC + 2% SBS mixture), structure S2 (surface with 3% NC mixture), and structure S1 (surface with conventional mixture).

First, comparing the structures S1 and S2, it could be observed that the use of the 3% NC mixture, in substitution to the conventional one, would represent an increase of 1.5 times in the fatigue lifespan of the surface. When the same comparison is carried out between the S1 and S4 structures, it is noted that the simultaneous addition of the polymer to the nanoclay modified mixture would enhance these gains to about 23 times. However, despite the high performance of the 3% NC + 2% SBS mixture and keeping the thicknesses of the layers, it is noted that the use of the SBS 60/85, in substitution to the mixture modified simultaneously by nanoclay and polymer, would double the fatigue lifespan of the surface. In this sense, looking for the equivalent performance of these two mixtures, it requires an increase in the thickness of the 3% NC + 2% SBS modified surface. Even so, considering a possibility of cost reduction from the substitution of the SBS 60/85 mixture by the 3% NC + 2% SBS (due to the lower polymer content), it is expected that this substitution, besides being technically feasible, could also be economically viable. This finding highlights the application potential of the 3% NC + 2% SBS mixture, which may present as a competitive alternative to the other mixtures studied.

4. Conclusions

According to the objective previously defined, this study enabled four different asphalt mixtures to be compared in terms of their rheological behavior, resistance to permanent deformation, and resistance to fatigue cracking. The mixtures were prepared with different binders as follows: conventional (CAP 50/70), modified with nanoclay (3% NC), modified with SBS polymer (SBS 60/85), and modified with both nanoclay and SBS (3% NC + 2% SBS).

Based on the results obtained in this study, the mixture modified with the binder 3% NC + 2% SBS showed a better rheological behavior and a higher resistance to permanent deformation. This mixture presented the highest values for the complex modulus and the lowest phase angles, which is considered an indication of higher resistance to rutting. In the permanent deformation tests, it showed the lowest percentage of rutting depth, thus confirming the predictions based on the rheological study.

On the other hand, considering the fatigue tests results, a better resistance was observed in the SBS 60/85 mixture. In this respect, the predictions made according to the rheological behavior were partially verified. Even so, on the next stage of this study, numerical simulations of pavement structures demonstrated the use of the 3% NC + 2% SBS surface as a replacement for the SBS 60/85 being technically viable (with thickness adjustments), and considering the possibility of reducing costs from this substitution, it could also be economically viable. In this sense, considering the costs of the materials used in the research, it is believed that the mixture modified by nanoargila and SBS (with lower content) would be cheaper than that modified by a higher SBS content. However, economic viability with specific comparisons should be verified in future studies.

Thus, it is concluded that the use of polymers science together with the application of nanotechnology can lead to great advances in the area of modified asphalt mixtures for use on roadways where a higher performance of the paving materials is required. Especially in relation to rutting, it is highlighted that this alternative has great potential to provide improved asphalt surfaces, mainly in regions of tropical climate and on highways submitted to slow heavy traffic.

Author Contributions: Conceptualization, G.C.C., G.T., J.V.S.d.M. and M.F.M.; methodology, G.C.C., G.T., J.V.S.d.M. and M.F.M.; formal analysis, G.C.C., G.T., J.V.S.d.M., M.F.M., L.P.T. and L.C.d.L.; investigation, G.C.C., G.T., J.V.S.d.M. and M.F.M.; writing—original draft preparation, G.C.C. and T.G.; writing—review and editing, G.C.C., G.T., J.V.S.d.M., M.F.M., L.P.T. and L.C.d.L.

Funding: This research was funded by NATIONAL COUNCIL FOR SCIENTIFIC AND TECHNOLOGICAL DEVELOPMENT (CNPq), grant number 130697/2015-0, and by ASPHALT TECHNOLOGY NETWORK/PETROBRAS, grant number 2014/00282-9.

Acknowledgments: The authors would like to thank the National Council for Scientific and Technological Development (CNPq), and the Asphalt Technology Network/Petrobras. We are also grateful to Greca Asfaltos S.A. (for supplying the binders CAP 50/70 and SBS 60/85) and to Kraton (for providing the SBS polymer).

Conflicts of Interest: The authors declare no conflict of interest.

References

1. Bernucci, L.B.; Motta, L.M.G.; Ceratti, J.A.P.; Soares, J.B. *Pavimentação Asfáltica: Formação Básica Para Engenheiros*; Petrobras ABEDA: Rio de Janeiro, Brazil, 2008.
2. Ashish, P.K.; Singh, D.; Bohm, S. Investigation on influence of nanoclay addition on rheological performance of asphalt binder. *Road Mater. Pavement Des.* **2016**, *18*, 1007–1026. [CrossRef]
3. Cavalcanti, L.S. Efeito de Alguns Modificadores de Ligante Na Vida de Fadiga e Deformação Permanente de Misturas Asfálticas. Master's Thesis, Programa de Pós-Graduação em Engenharia Civil, COPPE, Universidade Federal do Rio de Janeiro, Rio de Janeiro, Brazil, 2010.
4. Crucho, J.M.P.; Neves, J.M.C.; Capitão, S.D.; Picado-Santos, L.G. Mechanical performance of asphalt concrete modified with nanoparticles: Nanosilica, zero-valent iron and nanoclay. *Constr. Build. Mater.* **2018**, *181*, 309–318. [CrossRef]
5. Goh, S.W.; Akin, M.; You, Z.; Shi, X. Effect of deicing solutions on the tensile strength of micro or nano-modified asphalt mixture. *Constr. Build. Mater.* **2011**, *25*, 195–200. [CrossRef]
6. Iskender, E. Evaluation of mechanical properties of nano-clay modified asphalt mixtures. *Measurement* **2016**, *93*, 359–371. [CrossRef]
7. Jahromi, S.G.; Andalibizade, B.; Vossough, S. Engineering properties of nanoclay modified asphalt concrete mixtures. *Arab. J. Sci. Eng.* **2010**, *35*, 89–103.
8. Jahromi, S.G.; Khodaii, A. Effects of nanoclay on rheological properties of bitumen binder. *Constr. Build. Mater.* **2009**, *23*, 2894–2904. [CrossRef]
9. Leite, L.F.M.; Guerra, G.; Martins, A.; Cravo, M. Efeito de nanomodificadores no envelhecimento e susceptibilidade térmica de cimentos asfálticos. In Proceedings of the 41ª Reunião Anual de Pavimentação. Fortaleza, Ceará, Brazil, 3–5 November 2012.
10. You, Z.; Mills-Beale, J.; Foley, J.; Roy, S.; Odegard, G.; Dai, Q.; Goh, S.W. Nanoclay-modified asphalt materials preparation and characterization. *Constr. Build. Mater.* **2011**, *25*, 1072–1078. [CrossRef]
11. Yu, J.; Feng, P.; Zhang, H.; Wu, S. Effect of organo-montmorillonite on aging properties of asphalt. *Constr. Build. Mater.* **2009**, *23*, 2636–2640. [CrossRef]
12. Zare-Shahabadi, A.; Shokuhfar, A.; Ebrahimi-Nejad, S. Preparation and rheological characterization of asphalt binders reinforced with layered silicate nanoparticles. *Constr. Build. Mater.* **2010**, *24*, 1239–1244. [CrossRef]
13. Babagoli, R.; Mohammadi, R.; Ameri, M. The rheological behavior of bitumen and moisture susceptibility modified with SBS and nanoclay. *Pet. Sci. Technol.* **2017**, *35*, 1085–1090. [CrossRef]
14. Cai, L.; Shi, X.; Xue, J. Laboratory evaluation of composed modified asphalt binder and mixture containing nano-silica/rock asphalt/SBS. *Constr. Build. Mater.* **2018**, *172*, 204–211. [CrossRef]
15. Farias, L.G.A.T.; Leitinho, J.L.; Amoni, B.C.; Bastos, J.B.S.; Soares, J.B.; Soares, S.A.; Sant'ana, H.B. Effects of nanoclay and nanocomposites on bitumen rheological properties. *Constr. Build. Mater.* **2016**, *125*, 873–883. [CrossRef]
16. Galooyak, S.S.; Dabir, B.; Nazarbeygi, A.E.; Moeini, A. Rheological properties and storage stability of bitumen/SBS/montmorillonite composites. *Constr. Build. Mater.* **2010**, *24*, 300–307. [CrossRef]
17. Golestani, B.; Nam, B.H.; Nejad, F.M.; Fallah, S. Nanoclay application to asphalt concrete: Characterization of polymer and linear nanocomposite-modified asphalt binder and mixture. *Constr. Build. Mater.* **2015**, *91*, 32–38. [CrossRef]
18. Golestani, B.; Nejad, F.M.; Galooyak, S.S. Performance evaluation of linear and nonlinear nanocomposite modified asphalts. *Constr. Build. Mater.* **2012**, *35*, 197–203. [CrossRef]
19. Merusi, F.; Giuliani, F.; Filippi, S.; Polacco, G. A model combining structure and properties of 160/220 bituminous binder modified with polymer/clay nanocomposites. A rheological and morphological study. *Mater. Struct.* **2014**, *47*, 819–838. [CrossRef]
20. Pamplona, T.F.; Amoni, B.C.; Alencar, A.E.V.; Lima, A.P.D.; Ricardo, N.M.P.S.; Soares, J.B.; Soares, S.A. Asphalt binders modified by SBS and SBS/nanoclays: Effect on rheological properties. *J. Braz. Chem. Soc.* **2012**, *23*, 639–647. [CrossRef]
21. Zhang, H.; Su, M.; Zhao, S.; Zhang, Y.; Zhang, Z. High and low temperature properties of nano-particles/polymer modified asphalt. *Constr. Build. Mater.* **2016**, *114*, 323–332. [CrossRef]

22. Melo, J.V.S. Desenvolvimento E Estudo Do Comportamento Reológico E Desempenho Mecânico De Concretos Asfálticos Modificados Com Nanocompósitos. Ph.D. Thesis, Pós-Graduação Em Engenharia Civil, Universidade Federal De Santa Catarina, Florianópolis, Brazil, 2014.
23. Marcon, M.F. Estudo e Comparação do Desempenho Mecânico e Reológico Entre Concretos Asfálticos Modificados Por Polímero SBS, Borracha Moída de Pneu e Nanomateriais. Master's Thesis, Pós-Graduação em Engenharia Civil, Universidade Federal de Santa Catarina, Florianópolis, Brazil, 2016.
24. Carlesso, G.C. Estudo do Comportamento de Mistura Asfáltica Modificada Por Nanoargila e Polímero SBS. Master's Thesis, Pós-Graduação em Engenharia Civil, Universidade Federal de Santa Catarina, Florianópolis, Brazil, 2017.
25. ASTM–American Society for Testing Materials. *ASTM C 127: Standard Test Method for Density, Relative Density (Specific Gravity), and Absorption of Coarse Aggregate*; ASTM: West Conshohocken, PA, USA, 2012.
26. DNER–Departamento Nacional de Estradas de Rodagem. *DNER-ME 084: Agregado Miúdo–Determinação da Densidade Real*; Método de ensaio; Instituto de Pesquisas Rodoviárias: Rio de Janeiro, Brazil, 1995.
27. DNER–Departamento Nacional de Estradas de Rodagem. *DNER-ME 085: Material Finamente Pulverizado–Determinação da Massa Específica Real*; Método de ensaio; Instituto de Pesquisas Rodoviárias: Rio de Janeiro, Brazil, 1994.
28. ASTM–American Society for Testing Materials. *ASTM D 5821: Standard Test Method for Determining the Percentage of Fractured Particles in Coarse Aggregate*; ASTM: West Conshohocken, PA, USA, 2013.
29. ASTM–American Society for Testing Materials. *ASTM C 1252: Standard Test Methods for Uncompacted Void Content of Fine Aggregate (as Influenced by Particle Shape, Surface Texture, and Grading)*; ASTM: West Conshohocken, PA, USA, 2006.
30. ABNT–Associação Brasileira de Normas Técnicas. *ABNT NBR 6954: Lastro-Padrão–Determinação da Forma do Material*; ABNT: Rio de Janeiro, Brazil, 1989.
31. AASHTO–American Association of State Highway and Transportation. *AASHTO T 176: Standard Method of Test for Plastic Fines in Graded Aggregates and Soils by Use of the Sand Equivalent Test*; Test standard specifications for transportation materials and methods of sampling and testing; AASHTO: Washington, DC, USA, 2008.
32. ASTM–American Society for Testing Materials. *ASTM C 131: Standard Test Method for Resistance to Degradation of Small-Size Coarse Aggregate by Abrasion and Impact in the Los Angeles Machine*; ASTM: West Conshohocken, PA, USA, 2014.
33. ASTM–American Society for Testing Materials. *ASTM C 88: Standard Test Method for Soundness of Aggregates by Use of Sodium Sulfate or Magnesium Sulfate*; ASTM: West Conshohocken, PA, USA, 2013.
34. AASHTO–American Association of State Highway and Transportation. *AASHTO T 112: Standard Method of Test for Clay Lumps and Friable Particles in Aggregates*; Test standard specifications for transportation materials and methods of sampling and testing; AASHTO: Washington, DC, USA, 2012.
35. ASTM–American Society for Testing Materials. *ASTM D 5: Standard Test Method for Penetration of Bituminous Materials*; ASTM: West Conshohocken, PA, USA, 2013.
36. ASTM–American Society for Testing Materials. *ASTM D 36–Standard Test Method for Softening Point of Bitumen (Ring-and-Ball Apparatus)*; ASTM: West Conshohocken, PA, USA, 2014.
37. ASTM–American Society for Testing Materials. *ASTM D 4402–Standard Test Method for Viscosity Determination of Asphalt at Elevated Temperatures Using a Rotational Viscometer*; ASTM: West Conshohocken, PA, USA, 2013.
38. ABNT–Associação Brasileira de Normas Técnicas. *ABNT NBR 15086: Materiais Betuminosos–Determinação da Recuperação Elástica Pelo Ductilômetro*; ABNT: Rio de Janeiro, RJ, Brazil, 2006.
39. Martinho, F.C.G.; Farinha, J.P.S. An overview of the use of nanoclay modified bitumen in asphalt mixtures for enhanced flexible pavement performances. *Road Mater. Pavement Des.* **2017**, *20*, 1–31. [CrossRef]
40. ABNT–Associação Brasileira de Normas Técnicas. *ABNT NBR 15166: Asfalto Modificado–Ensaio de Separação de Fase*; ABNT: Rio de Janeiro, RJ, Brazil, 2004.
41. AASHTO–American Association of State Highway and Transportation. *AASHTO M 323: Standard Specification for Superpave Volumetric Mix Design*; Test standard specifications for transportation materials and methods of sampling and testing; AASHTO: Washington, DC, USA, 2013.
42. AASHTO–American Association of State Highway and Transportation. *AASHTO R 35: Standard Practice for Superpave Volumetric Design for Hot-Mix Asphalt (HMA)*; Test standard specifications for transportation materials and methods of sampling and testing; AASHTO: Washington, DC, USA, 2012.

43. AFNOR–Association Française de Normalisation. *AFNOR NF P 98-250-2–Essais Relatifs Aux Chaussées, Preparation des Mélanges Hydrocarbonés, Partie 2: Compactage des Plaques*; AFNOR: Paris, France, 1991.
44. CEN–European Committee for Standardization. *EN 12697-26–Bituminous Mixtures, Test Methods for Hot Mix Asphalt–Part 26: Stiffness*; CEN: Brussels, Belgium, 2004.
45. CEN–European Committee for Standardization. *EN 12697-22+A1–Mélanges Bitumineux-Méthodes D'essai Pour Mélange Hydrocarboné à Chaud-Partie 22: Essai D'orniérage*; CEN: Brussels, Belgium, 2007.
46. LCPC–Laboratoire Central des Ponts et Chaussées. *Manuel LPC d'aide à La Formulation des Enrobés à Chaud: Groupe de travail RST Formulation des enrobés à chaud*; LCPC: Paris, France, 2007.
47. COST 333. *Development of New Bituminous Pavement Design Method: Final Report of the Action*; European Cooperation in the field of Scientific and Technical Research; European Commission Directorate General Transport: Brussels, Belgium, 1999.
48. CEN–European Committee for Standardization. *EN 12697-24–Bituminous Mixtures, Test Methods for Hot Mix Asphalt–Part 24: Resistance to Fatigue*; CEN: Brussels, Belgium, 2004.
49. Chabot, A.; Chupin, O.; Deloffre, L.; Duhamel, D. Viscoroute 2.0: A tool for simulation of moving load effects on asphalt pavement. *Road Mater. Pavement Des.* **2010**, *11*, 227–250. [CrossRef]
50. *ANP–Agência Nacional De Petróleo, Gás e Biocombustível*; Resolução ANP n° 32ANP: Brasília DF, Brazil, 2010.

© 2019 by the authors. Licensee MDPI, Basel, Switzerland. This article is an open access article distributed under the terms and conditions of the Creative Commons Attribution (CC BY) license (http://creativecommons.org/licenses/by/4.0/).

Article

Microwave Healing Performance of Asphalt Mixture Containing Electric Arc Furnace (EAF) Slag and Graphene Nanoplatelets (GNPs)

Federico Gulisano [1,*], João Crucho [2], Juan Gallego [1] and Luis Picado-Santos [2]

[1] Departamento de Ingeniería del Transporte, Territorio y Urbanismo, Universidad Politécnica de Madrid, C/Profesor Aranguren 3, 28040 Madrid, Spain; juan.gallego@upm.es
[2] CERIS, Instituto Superior Técnico, Universidade de Lisboa, Av. Rovisco Pais, 1049-001 Lisboa, Portugal; joao.crucho@tecnico.ulisboa.pt (J.C.); luispicadosantos@tecnico.ulisboa.pt (L.P.-S.)
* Correspondence: federico.gulisano@upm.es

Received: 24 January 2020; Accepted: 18 February 2020; Published: 20 February 2020

Abstract: Pavement preventive maintenance is an important tool for extending the service life of the road pavements. Microwave heating seems to be a promising technology for this application, as bituminous materials have the potential to self-repair above a certain temperature. As ordinary asphalt mixture has low microwave absorbing properties, some additives should be used to improve the heating efficiency. In this paper, the effect of adding Electric Arc Furnace (EAF) slag and Graphene Nanoplatelets (GNPs) on the microwave heating and healing efficiency of asphalt mixtures was evaluated. Microwave heating efficiency was assessed by heating the specimens using several heating times. In addition, the electrical resistivity of the mixtures was measured to understand its possible relationship with the microwave heating process. Furthermore, the healing rates of the asphalt mixtures were assessed by repeated Indirect Tensile Strength (ITS) tests. The results obtained indicate that the additions of graphene and EAF slag can allow important savings, up to 50%, on the energy required to perform a good healing process.

Keywords: graphene nanoplatelets (GNPs); EAF steel slag; asphalt mixtures; microwave heating; self-healing

1. Introduction

Cracking is one of the most common signs of asphalt pavement deterioration, producing a reduction of the mechanical strength and durability of the road pavement over time [1] affecting driving comfort and safety [2].

Generally, the cracks observed at the pavement surface can be caused by two major mechanisms, the bottom-up cracks and the top-down cracks. The bottom-up cracks are initiated by tensile strains at the bottom of the asphalt layer and the top-down cracks are initiated by surface tensile and shear stresses, environmental effects and ageing [3]. Fortunately, the asphalt mixture is a self-healing material, and if enough energy is applied, original mechanical properties can be partially or nearly totally restored. Such technology can enable an important reduction in the consumption of natural resources, saving aggregates and bitumen that would be used in reconstruction or repair/maintenance actions in the road network. By extending the service life of the current pavement, it may occur an overall reduction of the maintenance interventions, thus saving the corresponding costs and CO_2 emissions, as well as minimizing the traffic disruptions caused by such actions [4].

From a molecular point of view, the self-healing phenomenon is due to the wetting and interdiffusion of material between the two faces of a microcrack to achieve properties of the original material [5]. Sun et al. developed a healing function of asphalt material based on molecular diffusion theory [6]. Healing activation energy was found to be a promising parameter for evaluating self-healing ability as if appreciable energy equal to or greater than the healing activation energy exists at the damage faces, the self-healing reaction will start. Molecular diffusion models can only be used for describing the microcracks healing, as in the case of macrocracks, the molecular interdiffusion cannot occur due to a wider gap between the faces. In this case, the capillary flow healing model can be used to describe the self-healing phenomenon. Garcia explained that above a specific temperature, the bitumen behaves as a Newtonian fluid, and can fill the cracks in a sort of capillarity flow [7]. Different types of bitumen exhibit different threshold temperatures for flow, depending on their rheological properties, usually ranging from 30 °C to 70 °C. The flow behavior index of the bitumen was found to be appropriate to characterize the threshold for the initial self-healing temperature of the bitumen [8–10].

The healing capability of asphalt mixtures depends on several internal and external factors [11]. As for the internal factors, bitumen properties have a strong influence on healing capability. Bitumen with lower flow behavior index not only need less energy for starting healing, but also produce better healing levels [10]. At the microscale level, chemical composition has a strong influence on the healing properties of asphalt [12,13]. Cheng et al. demonstrated the influence of the surface energy on the healing capability of bitumen [14]. According to the findings of these authors, the most efficient healers should have relatively lower Lifshitz–van der Waals components and higher acid–base components of surface energy. The amount of bitumen content in the mixture increases the healing capability of asphalt pavement [15]. Some volumetric properties also influence the self-healing properties [16], as well as the type of mixture. Garcia et al. found that porous asphalt mixtures heal faster than the dense mixture, and provide better healing levels [10]. Between the external factors, rest time, temperature and damage degree are the most influential [9,17]. Many researchers have focused on finding the optimal temperature to maximize the healing effect. If the temperature is too low, bitumen cannot flow through the cracks. However, if the temperature is too high, the healing level decreases, probably due to the expansion of the asphalt mixture, which could cause structural defects in the pavement [17–20].

In the field, due to the continuous traffic flow, usually, the rest periods are not long enough to allow the self-healing to occur, and the pavement temperature rarely reaches the temperature needed for flowing. For this reason, in the last years, researchers have studied several technologies to promote the self-healing process. An example is the capsule healing, in which capsules containing rejuvenator oil are mixed with the asphalt materials [21,22]. When crack damage appears next to the capsules, they open and release the oil. The bitumen will be rejuvenated and the life of the asphalt mixture extended [23]. Another type of technology consists of heating the pavement, through induction or microwave heating, in order to reduce the viscosity of the bitumen and heal the cracks.

In the case of induction heating, electrical currents are induced by adding conductive particles in the composition of the mixture, and the heat is generated by the Joule effect [24–26]. Several additives and respective dosages were studied in order to maximize the conductivity of the mixture, such as steel wool, steel fibers, graphite, carbon black and carbon fibers [24,27–30]. Another heating technique is microwave heating, that was found to be more effective than induction heating to heal cracks in asphalt roads [19]. Microwaves are electromagnetic waves with frequencies ranging from 300 MHz to 300 GHz, and wavelengths from 1 m to 1 mm. In industrial applications, the frequencies 915 MHz and 2.45 GHz are the most commonly used, but for special applications, the frequency 5.8 GHz is increasingly used [31]. When microwave radiation is applied, the polar molecules of the asphalt mixture attempt to line up (polarization) with the alternating electromagnetic field. The inability of this polarization to follow the extremely rapid reversals of the electromagnetic field generates random motion and inter-molecular friction that produces heat [32,33]. Microwave heating basically depends on the strength and the frequency of electromagnetic field, the dielectric properties of the matter, which represents the efficiency of material in absorbing microwave energy, the conductivity losses and some thermal properties [34].

Several authors studied the dielectric properties of asphalt mixture for different applications, such as deicing [35], density measurement [36], recycling [33], and maintenance purposes [34,37]. Since the microwave susceptibility of conventional asphalt mixture is low, some authors have added microwave absorbing additives to the mixture for healing purposes [38], such as steel wool fibers [1,19,39,40], steel slag [2,37,41], steel shaving [42], ferrite [17,18] and carbon nanotubes [43]. Trigos et al. proposed a classification of different aggregates frequently used for pavements construction in terms of microwave heating efficiency [44].

Steel slag is a byproduct of the steel production process and is widely used in road pavements, due to his excellent mechanical properties, in terms of roughness, shape, angularity, hardness, polishing and wear resistance [45]. From the environmental point of view, the use of steel slag allows to reduce the amount of material to dispose of, and therefore, the incorporation in asphalt mixtures permits to convert a waste material into a resource. Additionally, the inclusion of slag does not imply any additional cost because the cost of the slag is similar to the prices of natural aggregates [38]. However, only a few studies focused on its microwave absorbing properties and its use for healing purposes in asphalt mixtures. Liu et al. proposed a method to increase the content of ferric oxide of steel slag particles, improving the microwave heating efficiency of asphalt mixture [37]. Li et al. studied the influence of steel slag filler on the self-healing properties of the asphalt mixture using a fatigue-healing-fatigue test. The results show an enhancement in the healing properties of the mixture containing steel slag [2]. Phan et al. used coarse steel slag and steel wool fibers in the asphalt mixture and evaluated the healing rate through three-point bending test [41]. The main result was that the addition of 30% steel slag increased the healing properties of the asphalt mixture.

Recently, with the advent of nanotechnology, some nanomaterials were used in asphalt mixtures. Several mechanical properties can be improved by adding some nanomaterials, such as nanosilica, nanoclay and nanoiron [46]. Recently, carbon and graphene family nanomaterials have been used for asphalt modification [47], such as graphene nanoplatelets (GNPs). The addition of GNP in the mixture leads to an improvement in flexural strength at low temperatures, better performance at high temperatures [48] and easier compaction [49]. However, only a few studies focused on the microwave heating and healing efficiency of asphalt mixture containing graphene nanomaterials. Li et al. used graphene to improve the microwave heating and healing properties of bitumen [8]. The results showed that graphene provides benefits in terms of heating and healing performance.

The objective of this paper is to evaluate the effect of adding Electric Arc Furnace (EAF) slag and Graphene Nanoplatelets (GNPs) on the microwave heating and healing efficiency of asphalt mixtures. This research is the continuation of the study carried out by Gallego et al. [50], where preliminary results of the heating efficiency of these additives were obtained. The graphene nanoplatelets were incorporated as a binder additive, while the EAF slag was added as partial replacement of the natural aggregates. The asphalt mixtures heating efficiency was evaluated using the ratio °C/kWh/kg and, in addition, the electrical resistivity was measured to understand the effect of conductivity losses in the heat generation process. The healing efficiency was studied by applying microwave energy to damaged specimens and evaluating the healing recovery using the Indirect Tensile Strength (ITS) test.

2. Materials and Methods

2.1. Materials

A conventional dense asphalt mixture AC20 35/50 (EN 13108-1:2007) was the mixture type selected to conduct the experimental study. The mixture particle size distribution is presented in Table 1. Limestone aggregates, limestone filler, and 35/50 conventional bitumen were the materials chosen for the production of the mixtures. According to the bitumen specification, the temperatures of 165 °C and 155 °C were adopted for the asphalt mixture production and compaction, respectively. The bitumen content, 4.7% by total weight of the mixture, was previously determined using the Marshall method. The mixing process was conducted using a laboratory mixer (EN 12697-35:2016). Cylindrical specimens of 100 mm in diameter and approximately 63.5 mm in height were compacted using a Marshall hammer (EN 12697-30:2004) applying 75 blows on each side of the specimen.

Table 1. Grading curve of the asphalt mixtures.

Sieve (mm)	% Passing
22	100
16	83
8	56
4	42
2	34
0.5	19
0.063	5

Six types of asphalt mixtures were produced in this study: one conventional reference mixture (with no additives), two mixtures with graphene (with 1% and 2% dosage by mass of modified binder) and three mixtures with EAF slag (with 3%, 6% and 9% of aggregate replacement).

Graphene nanoplatelets, commercially designated as GRAPHENIT-XL, were used to make the asphalt mixtures susceptible to microwaves. Regarding its chemical composition, graphene is essentially carbon (96.41%) with traces of other elements, such as oxygen (1.05%), sulphur (0.48%), nitrogen (0.48%), hydrogen (0.07%) and others. The graphene presents a bulk density of 0.04 g/cm^3. Similarly to other nanomaterials, the nanoscale of the graphene platelets enables a high specific surface area, thus occupying considerably more volume than conventional macroscopic particles. Figure 1 presents a sample of 2.50 g of graphene nanoplatelets and, by the left side, for comparison, 2.50 g of conventional limestone filler (fraction passing sieve 0.063 mm).

Figure 1. Mass of 2.50 g of limestone filler under 0.063 mm (**left**) and graphene nanoplatelets (**right**).

The Graphene nanoplatelets were dispersed in the bitumen matrix by adding the nanomaterial to the bitumen heated at 160 °C and applying high-speed mechanical stirring (2000 rpm) during 60 min. Additional details about the mixing process can be found elsewhere [51]. After each bitumen modification (with 1% and 2% graphene), the asphalt mixtures were produced and compacted as described for the conventional mixture.

The EAF slag used in this study is produced in Spain, and its chemical composition is reported in Table 2. After the hydration process carried out by the producer, the free calcium oxide becomes almost zero. This procedure prevents expansion problems associated with the presence of CaO. Two fractions of slag were used as a replacement of limestone aggregates, 0.5/2 mm and 0.063/0.5 mm. Figure 2 presents a sample of 50 g of both fractions. Although many investigations report that the substitution of coarse aggregates improves the mechanical properties of the mixture [45], in this research only fine aggregates were replaced, because this provides more homogeneous heating throughout the mixture, as reported by other authors [52]. Volumetric replacement principle was used in order to take into account the different bulk density of slag and natural aggregates [45].

Table 2. Chemical composition of EAF slag.

Chemical Composition	%
Al_2O_3	8.81
CaO	24.28
Fe_2O_3	40.49
MgO	3.02
MnO	4.72
SiO_2	12.60
P_2O_5	0.36
Other substances	5.72

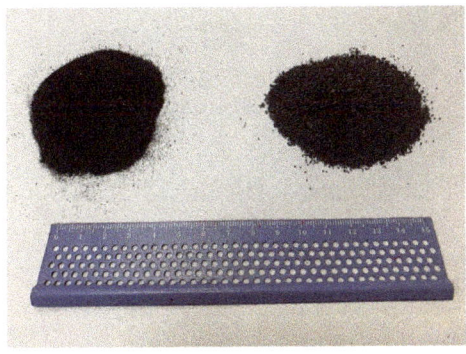

Figure 2. Mass of 50 g of slag, fraction 0.063/0.5 (**left**) and 0.5/2 mm (**right**).

2.2. Bulk Density of the Asphalt Mixtures

Bulk density of the asphalt mixtures (EN 12697-6:2012) was calculated in order to evaluate the physical properties of the mixtures with and without the addition of slag or graphene. Saturated surface dry (SSD) procedure was applied, in which the specimen is first saturated with water, and then its surface is blotted dry with a towel. The bulk density ρ_{bssd} is calculated as:

$$\rho_{bssd} = \frac{m_1}{m_3-m_2} \times \rho_w \quad (1)$$

where m_1 is the mass of the dry specimen in g; m_2 is the mass of the specimen in water in g; m_3 is the mass of the saturated surface-dried specimen in g and ρ_w is the density of the water at the test temperature.

2.3. Indirect Tensile Strength (ITS) Test

The effect of additives on the mechanical properties of the mixtures was evaluated with the Indirect Tensile Strength (ITS) test (EN 12697-23:2018). The selected test temperature was 15 °C. In the ITS test, a diametrical load was applied at a constant deformation rate of 50 ± 2 mm/min till the rupture of the specimen. Such loading produces tensile stress through the vertical diametral plane, as shown in Figure 3. To prevent excessive deformation of the specimens, the test was interrupted when the measured load dropped by 20% after the peak load. The Indirect Tensile Strength (ITS), in MPa, was calculated as:

$$ITS = \frac{2 \cdot P_{max}}{\pi \cdot d \cdot h} \quad (2)$$

where P_{max} is the peak load, in KN, d is the diameter of the specimen in mm and h is the height of the specimen, in mm.

Figure 3. Indirect Tensile Strength (ITS) test.

2.4. Microwave Heating

To heat the samples of asphalt mixture, a conventional microwave oven (with maximum power of 700 W and a frequency of 2.45 GHz) was used. In this study, the medium power level (350 W) was selected and used through all of the study. The heating efficiency of the mixtures was calculated as follows. The cylindrical specimens were cut into two pieces and then were conditioned at 25 °C for 2 h. Then, both pieces were placed in the microwave oven and heated for five heating times: 30 s, 60 s, 90 s, 120 s and 150 s. After each heating time, the two halves of the specimen were separated and an infrared thermometer was used to measure the internal temperature, as the average of eight randomly measurements, as shown in Figure 4. Additionally, the energy consumption during heating was measured with an electricity meter. Linear regression analysis was used to model the relationship between the internal temperature (°C) of the asphalt mixtures and the total energy consumption during the heating process (kWh/Kg).

Figure 4. Internal temperature measurement.

2.5. Electrical Resistivity Measurement

The electrical resistance of the asphalt mixture was measured with the two-probe method, by using a megohmmeter with 5 ranges (50 V–1000 V).

The asphalt specimen, with a height of about 4 cm, was placed between two copper plate electrodes with dimensions of 15 × 15 cm connected with the megohmmeter, as shown in Figure 5. In order to ensure perfect contact, graphene powder was used to fill the gaps between the plate electrodes and the specimens. The electrical resistivity was calculated applying the second Ohm's law:

$$\rho = \frac{R \cdot S}{l} \qquad (3)$$

where R is the electrical resistance of each specimen in Ω, S is the electrode-specimen contact area measured in m^2 and l is the thickness of the asphalt sample in m.

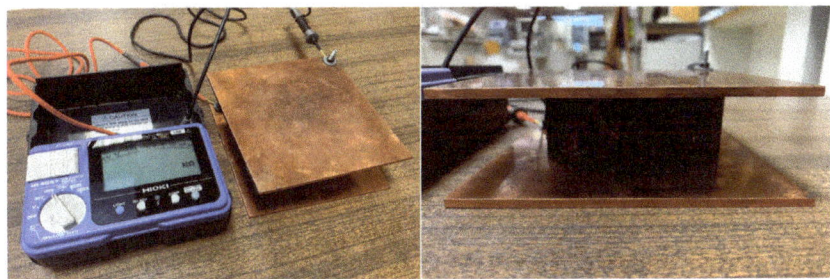

Figure 5. Electrical resistivity measurement.

2.6. Self-healing Test Procedure

The self-healing performance of the asphalt mixtures was evaluated as follow. First, each cylindrical specimen was tested under the Indirect Tensile Strength (ITS) test, according to Section 2.3. Then, the specimens were left at room temperature until they reached a temperature of 25 °C, and elastic rubber bands were used to tight the specimens before the heating treatment, as shown in Figure 6. A similar approach was used by other authors [53], which used a plastic collar to tight the specimen.

Figure 6. Simulation of in situ confining conditions of the asphalt pavements.

Preliminary tests conducted in the laboratory showed that the absence of the confining elastic rubber bands produces an enlargement of the crack during the heating process, due to the collapse of the mixture, that makes impossible the healing process. In situ, this cannot occur due to the lateral confinement of the asphalt mixture in the pavement. This phenomenon can be observed in Figure 7, where the broken specimen (Figure 7a) was heated without the inclusion of the rubber bands, causing an enlargement of the crack (Figure 7b).

Figure 7. The broken specimen before heating (**a**), and the enlargement of the crack after the heating without rubber bands (**b**).

The rubber bands were, therefore, used to approximately simulate the in situ confining conditions of the asphalt mixture in the pavement, and to obtain a more realistic measurement of the healing performances. Then, microwave radiation was applied. In order to evaluate the effect of temperature on the healing efficiency, the specimens were heated at different internal temperatures, 40 °C, 60 °C, 80 °C, 100 °C. Times required to reach these internal temperatures were found using the linear models obtained as described in Section 2.4. After the heating process, the specimens rested at 25 °C for 24 h, and then, after removing the rubber bands, Indirect Tensile Strength (ITS) test at 15 °C was repeated in order to evaluate the Healing Rate (HR):

$$\mathrm{HR}(\%) = \frac{(\mathrm{ITS}_{fin})}{(\mathrm{ITS}_{in})} \cdot 100 \quad (4)$$

where ITS_{fin} is the Indirect Tensile Strength of the sample after the healing process and ITS_{in} is the Indirect Tensile Strength of the sample initially tested.

The schematic representation of the methodology is reported in Figure 8.

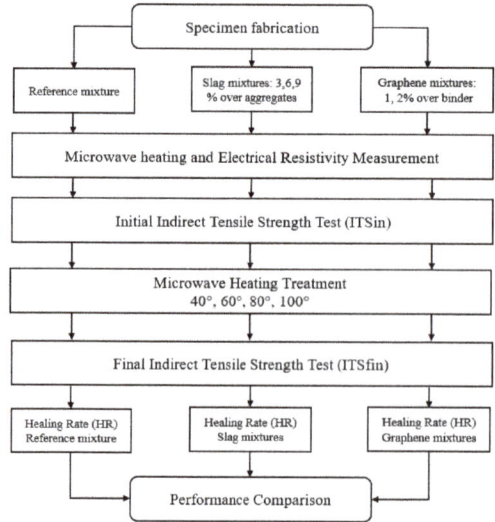

Figure 8. Flow chart of the healing process.

Furthermore, the analysis of variance (ANOVA) was performed to evaluate the effect of the temperature and the additive content on the healing rates of the asphalt mixtures.

3. Results

3.1. Influence of Slag and Graphene on the Physical and Mechanical Properties of the Asphalt Mixtures

The effect of additives on the bulk density of the asphalt mixtures is shown in Figure 9. The values are the average of 12 specimens, and the error bars represent the standard deviation. The reference mixture, without additives, has a bulk density of 2.452 g/cm^3. It can be observed that by adding slag to the mixture, the bulk density increases, until 2.481 g/cm^3, 2.503 g/cm^3 and 2.526 g/cm^3 for mixtures with 3%, 6% and 9% of slag, respectively. This trend is due to the higher specific gravity of the slag respect to natural aggregate, as reported also by other authors [45,54,55]. In contrast, graphene addition has the effect of decreasing the bulk density of the mixtures, until 2.429 g/cm^3 and 2.412 g/cm^3 for mixtures with 1% and 2% of graphene, respectively. This effect was probably due to the presence of graphene that increased the viscosity of the bitumen, as any powdered filler incorporated in the bitumen. Therefore, as the mixing and compaction temperatures were kept constant regardless of the content of graphene, for comparative purposes, the compaction was less effective when incorporating graphene.

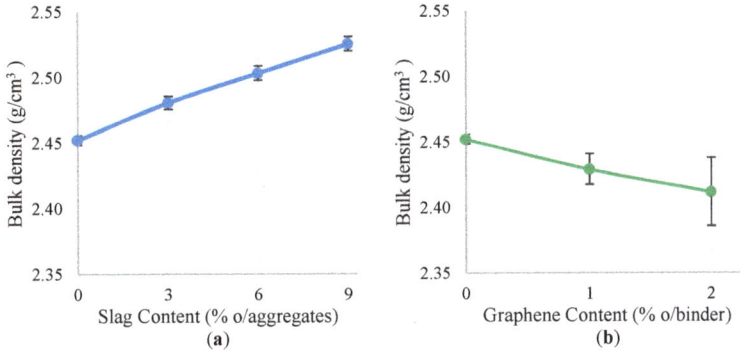

Figure 9. Effect of slag (a) and graphene (b) on the bulk density.

The effect of additives on the initial Indirect Tensile Strength (ITS_{in}) of the mixtures is shown in Figure 10. The values are the average of 12 specimens, and the error bars represent the standard deviation. It can be observed that the addition of slag or graphene has no important effect on the ITS_{in} of the mixtures.

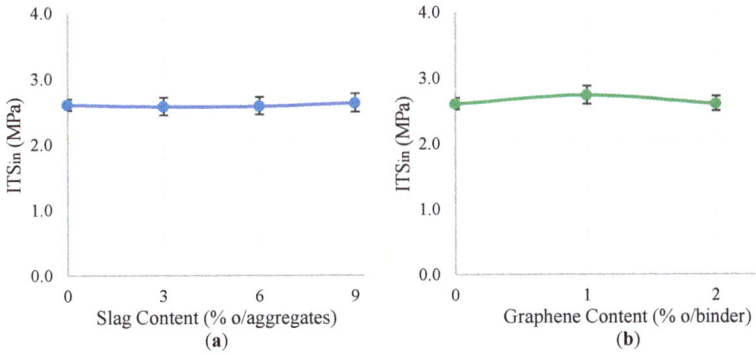

Figure 10. Effect of slag (a) and graphene (b) on the Indirect Tensile Strength ITS_{in}.

3.2. Influence of Slag and Graphene on the Heating Efficiency of the Asphalt Mixtures

The effect of adding slag or graphene to the mixture is an increase in the heating rates, as shown in Figure 11. The higher the amount of additive, the faster the temperature increase with energy. Even the mixture without additives can be heated by microwaves, although more energy, and consequently more heating time, must be applied to reach the same temperature. This means that microwave heating technique can also be used for existing pavements without additives, as also reported by other authors [56]. It can be observed in Figure 11 and Table 3 that the lineal models fit well the data, in terms of R^2. Nevertheless, it is interesting to analyze the effect of adding slag or graphene in terms of energy saving. The addition of 3 %, 6 % and 9 % (o/aggregates) of slag allows to save, respectively, 29%, 37% and 45% of the heating energy, respect to the ordinary asphalt mixture, while the addition of 1% and 2% (o/binder) of graphene allows to save 29% and 50% of the heating energy. This improvement of energy efficiency can produce several benefits in terms of CO_2 emissions and maintenance costs.

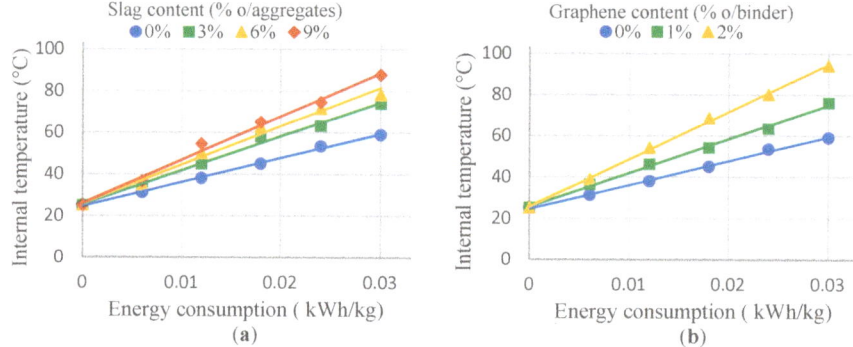

Figure 11. Effect of slag (**a**) and graphene (**b**) on the microwave heating consumption.

Table 3. Output of the linear regression models.

Additive	Content (%)	Linear Regression Equation	R^2	°C/ kWh/kg	°C/s
-	0	y = 1160x + 24.57	0.998	1160	0.232
Slag (% o/aggregates)	3	y = 1634x + 25.48	0.993	1634	0.327
	6	y = 1850x + 25.98	0.987	1850	0.370
	9	y = 2089x + 25.89	0.992	2089	0.418
Graphene (%o/binder)	1	y = 1638x + 25.60	0.996	1638	0.328
	2	y = 2301x + 25.58	0.998	2301	0.460

3.3. Influence of Slag and Graphene on the Electrical Resistivity of the Asphalt Mixtures

According to other studies [27–30], the electrical resistivity of the mixture slightly decreases with the additive content, until a critical value, called percolation threshold, where resistivity sharply decreases. The effect of adding slag or graphene on the electrical resistivity of the mixture is shown in Figure 12. In the case of slag addition, the percolation threshold is reached approximately between 6% and 9% of slag, when the electrical resistivity passes from 1.5×10^8 Ω·m to 1.7×10^6 Ω·m, corresponding to a reduction of 99%. Nevertheless, in the case of graphene, higher contents should be added to reach the percolation threshold. Comparing these results with the heating models (Figure 11), the contribution of the conductivity to the microwave heating can be analyzed. If the conductivity influenced the microwave heating, the sudden reduction of electrical resistivity (percolation threshold) would have led to a drastic increase in the heating rates. However, observing the heating curves of slag mixtures (Figure 11), the increase of the heating rate (°C/ kWh/kg) with the slag content is almost linear. This result can be referred to the fact that at microwaves frequencies, ranging from 300 MHz to 300 GHz, the conductivity contribution to heating is very low, and the heat is produced mostly by dipolar polarization rather than by the current created and the resulting Joule's effect. In this sense, the dielectric and thermal properties of the mixtures should be analyzed in future researches, in order to better understand the microwave heating phenomenon of asphalt mixtures and optimize the heating process.

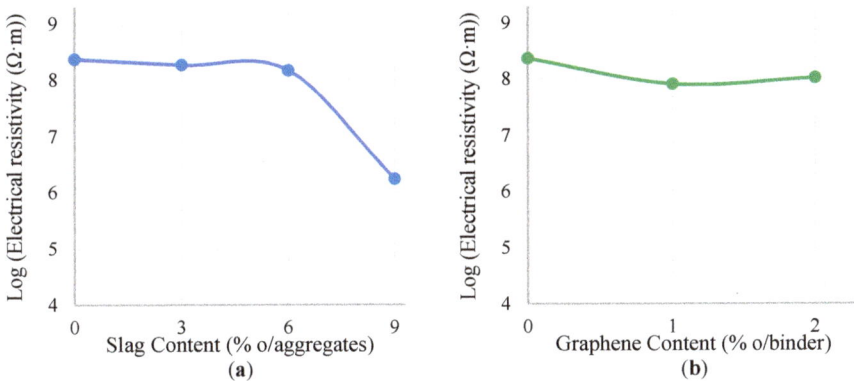

Figure 12. Effect of slag (**a**) and graphene (**b**) on the electrical resistivity.

3.4. Influence of Slag and Graphene on the Healing Properties of the Asphalt Mixtures

The results of the healing test are shown in Figure 13. The values represent the average Healing Rate (HR) of 3 samples, and the error bars represent the standard deviation. A one-way analysis of variance (ANOVA) was performed to evaluate the effect of the temperature on the healing rate of the asphalt mixtures. Normality and homogeneity of variances assumptions were checked. The analysis showed that the effect of the temperature was significant, $F(3,68) = 46.96$, p-value = 0.000. Post hoc comparisons using the Tukey HSD test indicated that all the means were significantly different from each other (p-value < 0.05). Therefore, the effect of the temperature was an increment in the healing rate of the asphalt mixtures. In fact, as described in Section 1, by increasing the temperature, the bitumen reduces its viscosity and flows through the open cracks easier, healing them. At 100 °C, for all the mixtures under study, the average HR is 67% (SD = 3.9), while the average HR for the mixtures heated at 40 °C is 44% (SD = 2.9). These results are consistent with those obtained by other authors [20,40,42].

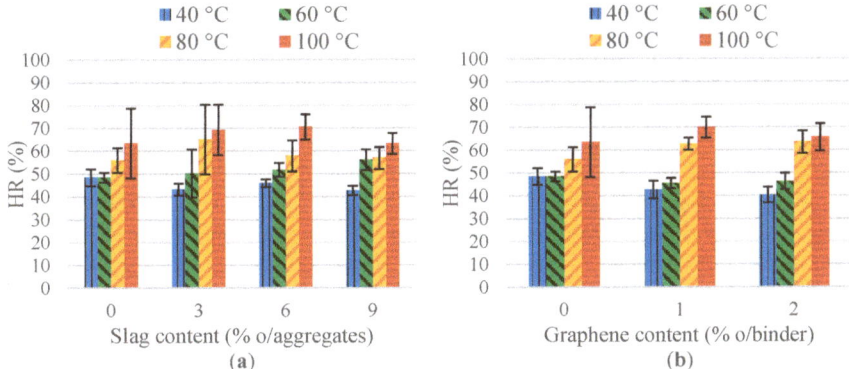

Figure 13. Effect of slag (**a**) and graphene (**b**) on the Healing Rate (HR)

A one-way analysis of variance (ANOVA) was performed to evaluate the effect of the additive content on the healing rates of the asphalt mixtures. Normality and homogeneity of variances assumptions were checked. The analysis showed that the effect of the additive content was not significant, $F(5,66) = 0.15$, p-value = 0.98. Therefore, the addition of slag or graphene did not produce substantial benefits in terms of the healing rate of the asphalt mixtures. However, the benefits in terms of energy savings are important. In this sense, Figure 14 shows the relationship between the total energy consumption during the heating and the healing rate of the mixtures. These curves should be interpreted as an indicator of the healing efficiency of the energy consumed by the microwave heating technique. The greater the slope of the curves, the higher the healing efficiency. In the case of slag, considerable benefits can be obtained even with the addition of 3% (o/aggregates), while with higher contents, no improvements are achieved. Similarly, in the case of graphene, the improvement of the healing efficiency is obtained with the addition of 2% (o/binder). For example, in order to obtain HR = 60%, the healing efficiency of the mixtures with 3% of slag and 2% of graphene is about double compared to the reference mixture. In this way, approximately half of the energy for pavements maintenance operations would be saved.

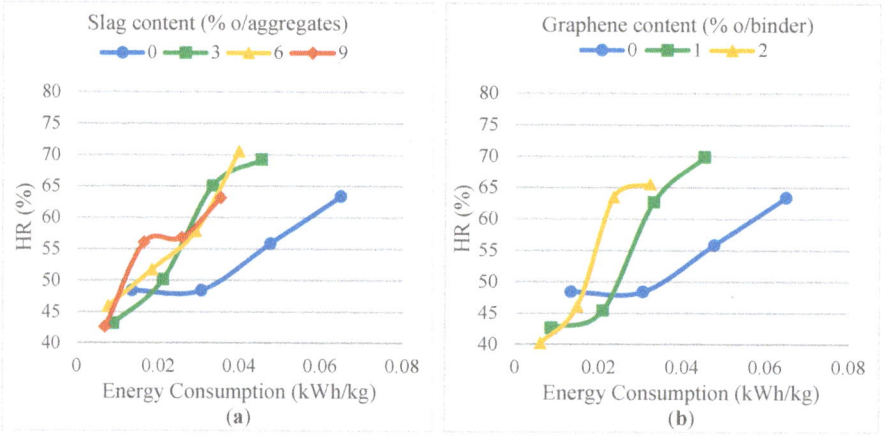

Figure 14. Effect of slag (**a**) and graphene (**b**) on the Healing Efficiency

4. Conclusions

In this paper, the effect of adding Electric Arc Furnace (EAF) slag or Graphene Nanoplatelets (GNPs) on the microwave heating and healing efficiency of asphalt mixtures was evaluated. The following conclusions can be drawn:

- It was found that the higher temperatures enhanced the healing performances of the asphalt mixtures, with and without the addition of slag or graphene.
- Although the asphalt mixture without additives can be heated with microwaves, both the slag and graphene allow saving approximately 50% of the energy during the heating process.
- The addition of slag or graphene does not seem to enhance the healing rates of the mixture. However, for the same healing rate, the addition of slag or graphene can halve the energy consumption during pavements maintenance operations. In accordance with the laboratory findings, the addition of 3% of slag (o/aggregates) or 2% of graphene (o/binder) is sufficient to obtain these savings of energy.
- Finally, it was observed that apparently, the contribution of the electrical conductivity to the microwave heating is low. Therefore, microwave heat generation could be attributed mostly to the

oscillating electromagnetic fields that excite molecules rather than to eventual electrical currents generated in the mixture and the resulting Joule's effect.

The use of GNPs in the asphalt mixtures is an innovative field, and the results of this work can be a starting point for further investigations in this area. In future research, other types of nanomaterials and the optimum dosage can be analyzed and compared with other traditional materials used in asphalt pavements, such as EAF slag.

Author Contributions: Conceptualization, J.G. and L.P.-S.; methodology, J.G. and L.P.-S.; validation, F.G and J.C.; formal analysis, F.G., J.C.; investigation, F.G., J.G. J.C., and L.P.-S.; resources, J.G. and L.P.-S.; writing—original draft preparation, F.G.; writing—review and editing, F.G., J.C., J.G. and L.P.-S.; supervision, J.G. and L.P.-S. All authors have read and agreed to the published version of the manuscript.

Funding: Fundación Agustín de Betancourt and Spanish Ministry of Education.

Acknowledgments: This investigation was possible thanks to Grant for PhD students by the Agustín of Betancourt Foundation and Grant n. PDI-18-0HXIUI-55-QZMWQL for Foreing Investigation Stays by the Spanish Ministry of Education.

Conflicts of Interest: The authors declare no conflict of interest.

References

1. Norambuena-Contreras, J.; Serpell, R.; Vidal, G.V.; Gonzalez, A.; Schlangen, E. Effect of fibres addition on the physical and mechanical properties of asphalt mixtures with crack-healing purposes by microwave radiation. *Constr. Build. Mater.* **2016**, *127*, 369–382. [CrossRef]
2. Li, C.; Wu, S.; Chen, Z.; Tao, G.; Xiao, Y. Enhanced heat release and self-healing properties of steel slag filler based asphalt materials under microwave irradiation. *Constr. Build. Mater.* **2018**, *193*, 32–41. [CrossRef]
3. Qiu, J. *Self Healing of Asphalt Mixtures: Towards a Better Understanding of the Mechanism*; Delft University of Technology: Delft, The Netherlands, 2012.
4. Tabaković, A.; Schlangen, E. Self-Healing Technology for Asphalt Pavements. In *Advances in Polymer Science*; Springer: New York, NY, USA, 2015; Volume 273, pp. 285–306. [CrossRef]
5. Little, D.N.; Bhasin, A. *Exploring Mechanism of Healing in Asphalt Mixtures and Quantifying Its Impact*; Springer: Dordrecht, The Netherlands, 2007; pp. 205–218.
6. Sun, D.; Lin, T.; Zhu, X.; Cao, L. Calculation and evaluation of activation energy as a self-healing indication of asphalt mastic. *Constr. Build. Mater.* **2015**, *95*, 431–436. [CrossRef]
7. García, Á. Self-healing of open cracks in asphalt mastic. *Fuel* **2012**, *93*, 264–272. [CrossRef]
8. Li, C.; Wu, S.; Chen, Z.; Tao, G.; Xiao, Y. Improved microwave heating and healing properties of bitumen by using nanometer microwave-absorbers. *Constr. Build. Mater.* **2018**, *189*, 757–767. [CrossRef]
9. Xiang, H.; He, Z.; Chen, L.; Zhu, H.; Wang, Z. Key Factors and Optimal Conditions for Self-Healing of Bituminous Binder. *J. Mater. Civ. Eng.* **2019**, *31*, 04019172. [CrossRef]
10. García, A.; Norambuena-Contreras, J.; Bueno, M.; Partl, M.N. Single and multiple healing of porous and dense asphalt concrete. *J. Intell. Mater. Syst. Struct.* **2015**, *26*, 425–433. [CrossRef]
11. Sun, D.; Sun, G.; Zhu, X.; Guarin, A.; Li, B.; Dai, Z.; Ling, J. A comprehensive review on self-healing of asphalt materials: Mechanism, model, characterization and enhancement. *Adv. Colloid Interface Sci.* **2018**, *256*, 65–93. [CrossRef]
12. Kim, Y.R.; Little, D.N.; Benson, F.C. Chemical and mechanical evaluation on healing mechanism of asphalt concrete. *J. Assoc. Asph. Paving Technol.* **1990**, *59*, 240–275.
13. Sun, D.; Yu, F.; Li, L.; Lin, T.; Zhu, X.Y. Effect of chemical composition and structure of asphalt binders on self-healing. *Constr. Build. Mater.* **2017**, *133*, 495–501. [CrossRef]
14. Cheng, D.; Little, D.N.; Lytton, R.L.; Holste, J.C. Surface Energy Measurement of Asphalt and Its Application to Predicting Fatigue and Healing in Asphalt Mixtures. *Transp. Res. Rec. J. Transp. Res. Board* **2002**, *1810*, 44–53. [CrossRef]
15. Molenaar, A.A.A. *Design of Flexible Pavements*; Delft University of Technology: Delft, The Netherlands, 2007.
16. Ayar, P.; Moreno-Navarro, F.; Rubio-Gámez, M.C. The healing capability of asphalt pavements: A state of the art review. *J. Clean. Prod.* **2016**, *113*, 28–40. [CrossRef]

17. Zhu, X.; Cai, Y.; Zhong, S.; Zhu, J.; Zhao, H. Self-healing efficiency of ferrite-filled asphalt mixture after microwave irradiation. *Constr. Build. Mater.* **2017**, *141*, 12–22. [CrossRef]
18. Zhu, X.; Ye, F.; Cai, Y.; Birgisson, B.; Lee, K. Self-healing properties of ferrite-filled open-graded friction course (OGFC) asphalt mixture after moisture damage. *J. Clean. Prod.* **2019**, *232*, 518–530. [CrossRef]
19. Norambuena-Contreras, J.; Garcia, A. Self-healing of asphalt mixture by microwave and induction heating. *Mater. Des.* **2016**, *106*, 404–414. [CrossRef]
20. Norambuena-Contreras, J.; Gonzalez, A.; Concha, J.L.; Gonzalez-Torre, I.; Schlangen, E. Effect of metallic waste addition on the electrical, thermophysical and microwave crack-healing properties of asphalt mixtures. *Constr. Build. Mater.* **2018**, *187*, 1039–1050. [CrossRef]
21. García, Á.; Schlangen, E.; van de Ven, M.; Sierra-Beltrán, G. Preparation of capsules containing rejuvenators for their use in asphalt concrete. *J. Hazard. Mater.* **2010**, *184*, 603–611. [CrossRef]
22. Garcia, A.; Schlangen, E.; van de Ven, M. Two Ways of Closing Cracks on Asphalt Concrete Pavements: Microcapsules and Induction Heating. *Key Eng. Mater.* **2009**, *417–418*, 573–576. [CrossRef]
23. Schlangen, E.; Sangadji, S. Addressing Infrastructure Durability and Sustainability by Self Healing Mechanisms—Recent Advances in Self Healing Concrete and Asphalt. *Procedia Eng.* **2013**, *54*, 39–57. [CrossRef]
24. Liu, Q.; Wu, S.; Schlangen, E. Induction heating of asphalt mastic for crack control. *Constr. Build. Mater.* **2013**, *41*, 345–351. [CrossRef]
25. Liu, Q.; García, Á.; Schlangen, E.; van de Ven, M. Induction healing of asphalt mastic and porous asphalt concrete. *Constr. Build. Mater.* **2011**, *25*, 3746–3752. [CrossRef]
26. Garcia, A.; Bueno, M.; Norambuena-Contreras, J.; Partl, M.N. Induction healing of dense asphalt concrete. *Constr. Build. Mater.* **2013**, *49*, 1–7. [CrossRef]
27. Wu, S.; Mo, L.; Shui, Z.; Chen, Z. Investigation of the conductivity of asphalt concrete containing conductive fillers. *Carbon N. Y.* **2005**, *43*, 1358–1363. [CrossRef]
28. García, Á.; Schlangen, E.; van de Ven, M.; Liu, Q. Electrical conductivity of asphalt mortar containing conductive fibers and fillers. *Constr. Build. Mater.* **2009**, *23*, 3175–3181. [CrossRef]
29. Arabzadeh, A.; Ceylan, H.; Kim, S.; Sassani, A.; Gopalakrishnan, K.; Mina, M. Electrically-conductive asphalt mastic: Temperature dependence and heating efficiency. *Mater. Des.* **2018**, *157*, 303–313. [CrossRef]
30. Wang, H.; Yang, J.; Liao, H.; Chen, X. Electrical and mechanical properties of asphalt concrete containing conductive fibers and fillers. *Constr. Build. Mater.* **2016**, *122*, 184–190. [CrossRef]
31. Von Starck, A.; Muhlbauer, A.; Kramer, C. *Handbook of Thermoprocessing Technologies: Fundamentals, Processes, Components, Safety*; Vulkan-Verlag: Essen, Germany, 2005.
32. Metaxas, A.; Meredith, R. *Industrial Microwave Heating*; The Institution of Electrical Engineers: London, UK, 1983.
33. Benedetto, A.; Calvi, A. A pilot study on microwave heating for production and recycling of road pavement materials. *Constr. Build. Mater.* **2013**, *44*, 351–359. [CrossRef]
34. Wang, H.; Zhang, Y.; Zhang, Y.; Feng, S.; Lu, G.; Cao, L. Laboratory and Numerical Investigation of Microwave Heating Properties of Asphalt Mixture. *Materials* **2019**, *12*, 146. [CrossRef] [PubMed]
35. Ding, L.; Wang, X.; Zhang, W.; Wang, S.; Zhao, J.; Li, Y. Microwave Deicing Efficiency: Study on the Difference between Microwave Frequencies and Road Structure Materials. *Appl. Sci.* **2018**, *8*, 2360. [CrossRef]
36. Jaselskis, E.J.; Grigas, J.; Brilingas, A. Dielectric Properties of Asphalt Pavement. *J. Mater. Civ. Eng.* **2003**, *15*, 427–434. [CrossRef]
37. Liu, W.; Miao, P.; Wang, S.-Y. Increasing Microwave Heating Efficiency of Asphalt-Coated Aggregates Mixed with Modified Steel Slag Particles. *J. Mater. Civ. Eng.* **2017**, *29*, 04017171. [CrossRef]
38. Gallego, J.; del Val, M.A.; Contreras, V.; Páez, A.; Páez, A. Use of additives to improve the capacity of bituminous mixtures to be heated by means of microwaves. *Mater. Construcción* **2017**, *67*, 110. [CrossRef]
39. Gallego, J.; del Val, M.A.; Contreras, V.; Paez, A. Heating asphalt mixtures with microwaves to promote self-healing. *Constr. Build. Mater.* **2013**, *42*, 1–4. [CrossRef]
40. Gonzalez, A.; Norambuena-Contreras, J.; Storey, L.; Schlangen, E. Effect of RAP and fibers addition on asphalt mixtures with self-healing properties gained by microwave radiation heating. *Constr. Build. Mater.* **2018**, *159*, 164–174. [CrossRef]
41. Phan, T.M.; Park, D.-W.; Le, T.H.M. Crack healing performance of hot mix asphalt containing steel slag by microwaves heating. *Constr. Build. Mater.* **2018**, *180*, 503–511. [CrossRef]

42. Gonzalez, A.; Norambuena-Contreras, J.; Storey, L.; Schlangen, E. Self-healing properties of recycled asphalt mixtures containing metal waste: An approach through microwave radiation heating. *J. Environ. Manag.* **2018**, *214*, 242–251. [CrossRef]
43. Pérez, I.; Agzenai, Y.; Pozuelo, J.; Sanz, J.; Baselga, J.; García, A.; Pérez, V. Self-healing of asphalt mixes, containing conductive modified bitumen, using microwave heating. In Proceedings of the 6th Eurasphalt & Eurobitume Congress, Prague, Czech Republic, 1–3 June 2016. [CrossRef]
44. Trigos, L.; Gallego, J.; Escavy, J.I. Heating potential of aggregates in asphalt mixtures exposed to microwaves radiation. *Constr. Build. Mater.* **2020**, *230*, 117035. [CrossRef]
45. Skaf, M.; Manso, J.M.; Aragón, Á.; Fuente-Alonso, J.A.; Ortega-López, V. EAF slag in asphalt mixes: A brief review of its possible re-use. *Resour. Conserv. Recycl.* **2017**, *120*, 176–185. [CrossRef]
46. Crucho, J.; Picado-Santos, L.; Neves, J.; Capitão, S. A Review of Nanomaterials' Effect on Mechanical Performance and Aging of Asphalt Mixtures. *Appl. Sci.* **2019**, *9*, 3657. [CrossRef]
47. Wu, S.; Tahri, O. State-of-art carbon and graphene family nanomaterials for asphalt modification. *Road Mater. Pavement Des.* **2019**, 1–22. [CrossRef]
48. Hafeez, M.; Ahmad, N.; Kamal, M.A.; Rafi, J.; Zaidi, S.B.A.; Nasir, M.A. Experimental Investigation into the Structural and Functional Performance of Graphene Nano-Platelet (GNP)-Doped Asphalt. *Appl. Sci.* **2019**, *9*, 686. [CrossRef]
49. Le, J.-L.; Marasteanu, M.O.; Turos, M. Mechanical and compaction properties of graphite nanoplatelet-modified asphalt binders and mixtures. *Road Mater. Pavement Des.* **2019**, 1–16. [CrossRef]
50. Gallego, J.; Gulisano, F.; Picado, L.; Crucho, J. Optimizing asphalt mixtures to be heated by microwave. In Proceedings of the 17th International Conference on Microwave and High Frequency Heating, Valencia, Spain, 9–12 September 2019. [CrossRef]
51. Crucho, J.M.L.; Neves, J.M.C.d.; Capitão, S.D.; de Picado-Santos, L.G. Mechanical performance of asphalt concrete modified with nanoparticles: Nanosilica, zero-valent iron and nanoclay. *Constr. Build. Mater.* **2018**, *181*, 309–318. [CrossRef]
52. Gao, J.; Sha, A.; Wang, Z.; Tong, Z.; Liu, Z. Utilization of steel slag as aggregate in asphalt mixtures for microwave deicing. *J. Clean. Prod.* **2017**, *152*, 429–442. [CrossRef]
53. Tabaković, A.; O'Prey, D.; McKenna, D.; Woodward, D. Microwave self-healing technology as airfield porous asphalt friction course repair and maintenance system. *Case Stud. Constr. Mater.* **2019**, *10*, e00233. [CrossRef]
54. Magadi, K.L.; Anirudh, N.; Mallesh, K.M. Evaluation of Bituminous Concrete Mixture Properties with Steel Slag. *Transp. Res. Procedia* **2016**, *17*, 174–183. [CrossRef]
55. Neves, J.; Crucho, J.; Santos, L.P.; Martinho, F. The influence of processed steel slag on the performance of a bituminous mixture. In Proceedings of the Ninth International Conference on Bearing Capacity of Roads, Railways and Airfields, Trondhein, Norway, 25–27 June 2013; pp. 617–625.
56. Sun, Y.; Wu, S.; Liu, Q.; Zeng, W.; Chen, Z.; Ye, Q.; Pan, P. Self-healing performance of asphalt mixtures through heating fibers or aggregate. *Constr. Build. Mater.* **2017**, *150*, 673–680. [CrossRef]

© 2020 by the authors. Licensee MDPI, Basel, Switzerland. This article is an open access article distributed under the terms and conditions of the Creative Commons Attribution (CC BY) license (http://creativecommons.org/licenses/by/4.0/).

Article

Experimental Study on Photocatalytic Effect of Nano TiO$_2$ Epoxy Emulsified Asphalt Mixture

Ming Huang * and Xuejun Wen

Shanghai Municipal Engineering Design Institute (Group) Co., Ltd., Shanghai 200092, China; 13719145727@163.com
* Correspondence: huangming@tongji.edu.cn

Received: 14 May 2019; Accepted: 13 June 2019; Published: 17 June 2019

Featured Application: A new emulsified asphalt mixture, to which a specially applicable epoxy curing system was added, was used in this study; four key influence factors on the photocatalytic effect were investigated to guide application of TiO$_2$ to asphalt pavements; and average illumination of underground road surface as a design index for xenon lamp lighting systems was proposed.

Abstract: The two major problems that have plagued urban underground roads since their introduction are the harmful emissions caused by hot mix paving and vehicle exhaust accumulation during operation. In order to solve these two problems at the same time, a new asphalt mixture degrading automobile exhaust, which has the advantage of cold mix and cold-application, was presented and studied. A considerable amount of research shows that the use of titanium dioxide (TiO$_2$) for pavements has received considerable attention in recent years to improve air quality near large metropolitan areas. However, the proper method of applying TiO$_2$ to asphalt pavements is still unclear. The new mixture presented in this article contains epoxy emulsified asphalt as the binder; therefore, how to apply TiO$_2$ in the special asphalt mixture proves to be the main focus. By experimental design, four influence factors on the photocatalytic effect, which are the nano-TiO$_2$ particle sizes, dosage, degradation time, and light intensity, have been investigated. The experimental results showed that the 5-nm particle size of TiO$_2$ is better than 10–15 nm for exhaust gas degradation, especially for HC and NO; with an increase in the amount of photocatalytic material, the degradation of CO and CO$_2$ in the exhaust gas did not increase obviously, while the degradation effects of HC and NO were remarkable; in the 4-h time extended degradation test, the experimental data show that the extended time has little effect on the degradation rate of CO$_2$ and CO, and the general trend is that the degradation of exhaust became significant with the extension of time; while setting a 2-h NO degradation rate as an indicator, to make the index more than 50% or 25%, the average illumination of the road surface cannot be less than 60 lx or 40 lx.

Keywords: nano titanium dioxide; epoxy emulsified asphalt; photocatalysis; exhaust gas degradation

1. Introduction

Urban underground roads have initiated a new and convenient way for rapid growth of vehicle flows in metropolitan areas. They will play a major role in changing urban traffic conditions, reducing noise and the destruction of the urban three-dimensional space. However, they have a fundamental problem, that is, space is relative airtight, which will cause two problems in both the construction and operation periods [1,2]. Firstly, heavy emissions accumulate in underground roads and construction air conditions deteriorate. Secondly, in later stages of operation, there will be higher concentrations of automobile exhaust and more danger to the health of the citizens surrounding the air vents [3].

In order to solve the emission problem in construction, current researchers mostly advocate the use of warm mixing technology. This technology can usually bring the temperature of paving asphalt mixture from 170–190 °C down to 135–150 °C [4,5], which can slightly reduce the emissions in construction.

In order to solve the vehicle exhaust accumulation problem in the stage of operation, in recent years, researchers in road work have considered vehicle carrier-road pavement materials and developed many new degradable automobile exhaust pavement materials [6–8]. These photocatalysts are dissolved in a solution and sprayed on the surface of the road to be exposed to automobile exhaust and sunlight. However, the durability of these methods is insufficient because of the thin structure thickness [9]. The photocatalyst cannot stay on the road for a long time, and the road will soon lose the function of degrading the tail gas.

Hence, a high-performance nano titanium dioxide epoxy emulsified asphalt mixture, instead of a solution, has been introduced as a new pavement material to solve the two problems simultaneously. In this study, it is tentatively applied to a surface wear layer with a thickness of 1–3 cm. The objective of the current research was to investigate the effects of different influencing factors on asphalt pavement degradation exhaust, taking into account the nano material's particle size, dosage in the binder, degradation time, and light intensity [10–12]. The findings can be seen as important reference indexes in the production of nano titanium dioxide epoxy emulsified asphalt.

2. Materials and Mixture Design

2.1. TiO$_2$ Powder

HC, NO and CO compounds can be transformed into salt and water by nano-TiO$_2$ under photocatalysis that is an irradiation by a light source with a wavelength less than 387.5 nm [7,8]. Therefore, we needed to select a scheme to make titanium dioxide more in contact with air and ultraviolet light in the design. In this study, Anatase phase nano titanium dioxide was used as it has the best degradation effect among several known phases [13]. The size range of anatase titanium dioxide is large. In this study, 5 nm and 10–15 nm levels were selected. Their specific surface areas were 240 and 60–100 respectively, purities are both 99%.

2.2. Binder

The binder of epoxy asphalt was divided into two parts, Part A and Part B. Part A was epoxy resin, and Bisphenol A epoxy resin (type E-51) was chosen in this study [14]. Part B was a mixture of emulsified asphalt, titanium dioxide (TiO$_2$) powder, and the curing system. Table 1 presents the basic properties of Part B, with the test method according to ASTM specifications [15]. The basic properties of E-51 epoxy resin are shown in Table 2. In addition, the curing system in Part B comprises an amine curing agent, compatibilizer, and additives [16].

Table 1. Basic properties of Part B.

	Criterion	Unit	Detecting Result	Specification	Specification
	Residue by sieve test (1.18 mm)	%	0.01	≤0.1	T 0652
	Particle charge		Positive (+)	Positive (+)	T 0653
	Engler viscosity, E25	-	-	3–30	T 0622
	Residue by distillation	%	62.1	≥60	T 0651
	penetration (100 g, 25 °C, 5 s)	0.1 mm	67.6	40–100	T 0604
Test on residue	Softening point (R/B)	°C	59.2	≥57	T 0606
from distillation	Ductility (5 °C)	cm	>100	≥20	T 0605
	Solubility (trichloroethylene)	%	99.5	≥97.5	T 0607
Storage stability	1 d	%	0.2	≤1	T 0655
	5 d	%	2.3	≤5	

Table 2. Main chemical properties of E-51 type epoxy resin.

Chemical Composition	Viscosity (mPa·s)	Epoxy Equivalent(g/eq)	Density (23 °C) (g/cm^3)
2,2-bis(4-(2,3-epoxpropyloxy(phenyl)propane	11,000–14,000	211–290	≤1.10

2.3. Mixture Design

All aggregates used in this study were basalt because such aggregates produced by most ore fields exhibit a better shape and strength than others. The filler was limestone, and the vast majority of mineral fillers are made of limestone mainly because limestone powder combines well with asphalt [16], which can produce an effect similar to that of asphalt mastic, thereby effectively reducing the bleeding. The test results of the basic properties of the aggregates are summarized in Tables 3 and 4. Each aggregate was studied separately to fulfill the requirements of the material specifications in China [17,18].

Table 3. Aggregate gravity.

Size (mm)	Mineral Powder	0~3	3~5	5~10	10~13	13~19
gravity	2.788	2.835	2.866	2.875	2.903	2.909

Table 4. Property index of aggregates.

Test Index		Basalt Aggregate	Standard Requirement (JTJ F40-2004) [12]
Crushed stone value (%)		23.2	28
Weared stone value (Los Angeles) (%)		21.3	30
Content of flat particle (%) (size between 4.75 and 13.2 mm)		10.3	20
Sand equivalent value (size 2.36 mm) (%)		91.0	60
Angularity (%)	size between 2.36 and 4.75 mm	30	30
	Size 2.36 mm	49.2	

Table 5 presents the gradation of AC-13, which was referenced from the Technical Specifications [17].

Table 5. Gradation of asphalt mixture used in test.

Size (mm)	Grade	16.0	13.2	9.5	4.75	2.36	1.18	0.6	0.3	0.15	0.075
Passing rate (%)	AC-13	100	93	77	54	35	22	17	10	8	6

The optimum asphalt contents for the different modified asphalt mixtures have been determined using the Marshall mixture design (optimum asphalt content is 4.2% and the average of air void is 4.2%). Then, all of the above materials were supplied to produce all of the asphalt mixture specimens tested, in 30 × 30 cm rutting test form, using the Marshall design [18].

3. Experimental Design

3.1. Preparation of Carriers and Mixtures

The process of adding nano-titanium dioxide into the carrier of asphalt mixture is described as follows: Titanium dioxide is first dispersed in water, and eventually fused with epoxy emulsified asphalt, the mixture and specimens required for the test were prepared by adding aggregate. The process is shown in Figure 1 (part of the process and data are patented technology and appropriate concealments have been made).

Figure 1. Preparation process of nano TiO$_2$ epoxy emulsified asphalt and mixtures.

3.2. Optical Parameters of Laboratory Simulated Light Source

Titanium dioxide needed to be excited under a certain condition of illumination. Xenon lamps which are similar to the solar spectrum can be chosen as a light source. They have 400 nm–315 nm–280 nm wavelengths with a width of 3.10 to 4.43 eV ultraviolet light. The xenon lamp (power 25 W, the luminous flux 3200 lm), which can generate excited electrons and holes, was used as a safe photocatalytic light source [19]. The xenon lamp and the rutting test specimen were placed in the sealed tank of the automobile exhaust analyzer, as shown in Figure 2. For the calculation of the spatial distribution of light intensity, luminaire efficiency, luminance distribution, and shading angle, etc., the spatial arrangement is simplified as a mathematical geometric model (see Figures 3–5 below). The experiments were carried out at nighttime to simulate the dark environment of an underground road.

Three lighting arrangement schemes were used to obtain different photocatalytic reaction effects.

Scheme A: The number of xenon lamps: 1, position, top, height H = 0.2 m.

Scheme B: The number of xenon lamps: 2, position, both sides, height H = 0.2 m, elevation angle = 30°.

Scheme C: The number of xenon lamp: 3, position, both sides + top, height H = 0.2 m, elevation angle = 30°.

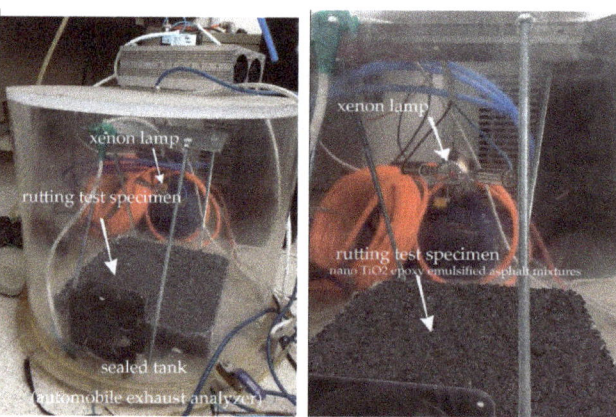

Figure 2. Rutting specimen and light source arrangement.

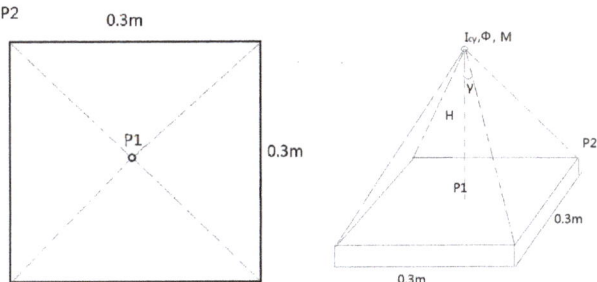

Figure 3. Illumination geometric model of Scheme A.

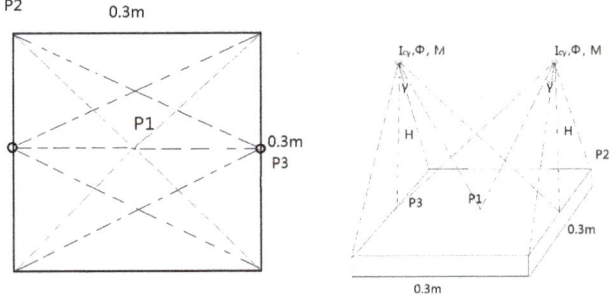

Figure 4. Illumination geometric model of Scheme B.

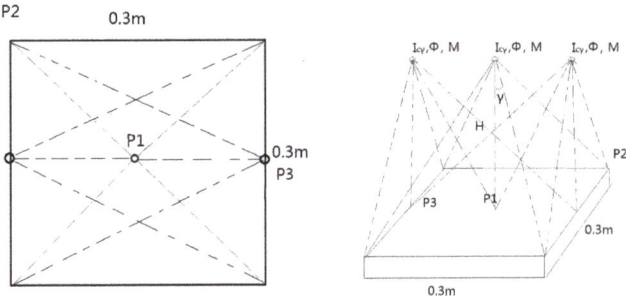

Figure 5. Illumination geometric model of Scheme C.

As shown in Figures 2–4, P1 is the maximum point of illumination, P2 is the weakest one, and P3 is the midpoint of the edge. The illumination value needs to be calculated constantly.

According to Reference [20], the illumination of the rutting panel is:

$$E_{pi} = \frac{I_{c\gamma}}{H^2} \cdot cos^3\gamma \frac{\Phi}{1000} M \tag{1}$$

In the formula,

E_{pi} –illumination of point P generated by the particular lamp (lx);

γ –light incident angle of point P from the particular lamp (°);

$I_{c\gamma}$ –light intensity value of point P (cd);

M –maintenance correction coefficient of the particular lamp, usually 0.6–0.7;

Φ –rated luminous flux of the particular lamp (lm);

H–the height of lamp light center to the road surface (m).

The single light source illuminance is calculated by Equation (1), while Equation (2) calculates the total illumination generated by several light sources.

$$E_p = \sum_{i=1}^{n} E_{pi} \qquad (2)$$

According to Equations (1) and (2), total illuminance of P1, P2, and P3 are calculated respectively. The calculation results are shown in Table 6 (unit, lx):

Table 6. Illuminance value of P1, P2, and P3 (unit, lx).

Typical Point \ Scheme	A	B	C
P1	36.22	48.31	84.53
P2	27.23	34.23	61.46
P3	-	40.56	40.56
Average of Luminosity	31.73	41.03	62.18

According to the illuminance design of conventional underground roads, luminosity is about 40–50 lx [20], Scheme B is consequently closer to the actual situation. In the following experimental studies, the xenon lamp model of Scheme B will be adopted.

3.3. Influence Factors

In this section, the impact of different factors on the degradation of vehicle exhaust will be studied, in order to determine the engineering design instructions. There are 4 influencing factors proposed, which are particle size of nano-TiO_2, dosage of nano-TiO_2, the duration and the illuminance of light [21,22]. Influence factors and related experimental design at different levels are shown in Table 7, and the parallel experiments are conducted three times in each grade. If the coefficient of variation of the result is greater than 10%, it will be removed, and more experiments will be carried out until three valid datasets were obtained. The whole experiment was carried out in an opaque closed room. The xenon lamp specified in the test is used for lighting.

Table 7. Experimental design of 4 influence factors.

Influence Factor	Grade					
Particle size of nanoscale titanium dioxide	5 nm				10–15 nm	
Dosage of nano TiO_2	10%		20%		30%	
Duration	2 h	3 h	4 h	6 h	10 h	24 h
Luminosity	31.73 lx		41.03 lx		62.18 lx	

4. Results and Discussion

4.1. Comparative Study on Two Particle Sizes

The 5 nm and 10–15 nm nano TiO_2 particles were tested for a two-hour test. The results are shown in Figure 6.

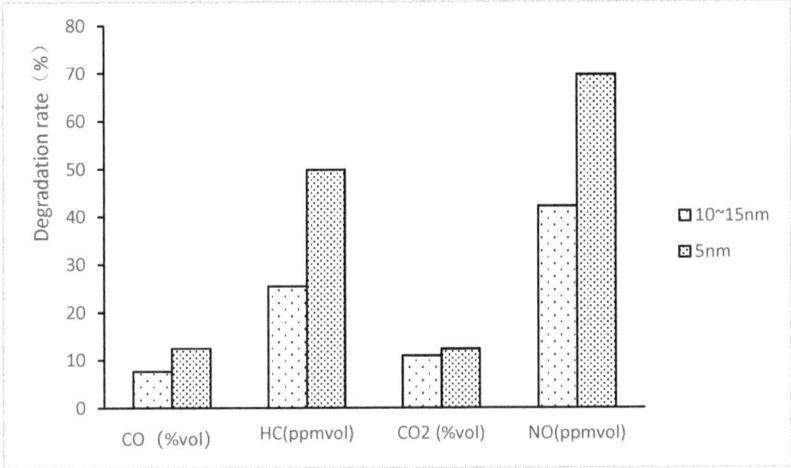

Figure 6. The comparison chart of exhaust degradation different particle size mixture.

With the smaller particle size of 5 nm, the degradation effect of exhaust is better than the 10–15 nm size, especially for HC and NO's degradation. The rest results will be presented for nano-TiO_2 in particles size of 5 nm.

4.2. Different Nano-TiO_2 Dosage

Nano-TiO_2 emulsion (content of 5%) was added during the asphalt emulsification process. Among the emulsions, solid content nano-TiO_2 emulsion dosages were 10%, 20% and 30% (a higher content of solid nano-TiO_2 cannot be dispersed, and a lower dosage will lead to less photocatalytic effect). This section mainly investigated the exhaust degradation of different dosages of 5 nm nano-TiO_2 added in the mixture. The test results of two hours are shown in Figure 7.

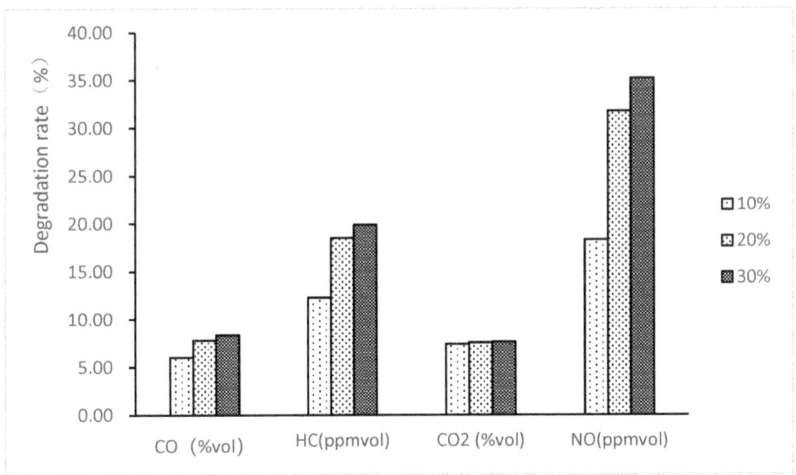

Figure 7. Exhaust degradation effect in different content of nano-TiO_2.

Through the analysis of experimental data, we can conclude that with the increase of the amount of nano-TiO_2, the degradation rate of exhaust gas of asphalt mixture tended to increase. However, the degradation of CO and CO₂ in the exhaust gas was not obvious. With the increase of the amount of nano-TiO$_2$, the degradation rate curves of the two were almost flat. With the amount of nano-TiO$_2$ increased from 10% to 20%, the degradation rate of NO and HC greatly increased. Relatively, with the content of nano-TiO$_2$ increased from 20% to 30%, the increase rate decreased.

4.3. Degradation Effect Changes with the Length of Degradation Time

According to the properties of nano-TiO$_2$ photocatalytic materials, nano-TiO$_2$ will not decrease in the chemical reactions. In this section, as the degradation reaction time is prolonged, the main study target is on the changes of degradation performance on exhaust gas. The exhaust degradation test sustained 4 h. The xenon lamp configuration is Scheme B, and the particle size is 5 nm. The test results are shown in Figure 8.

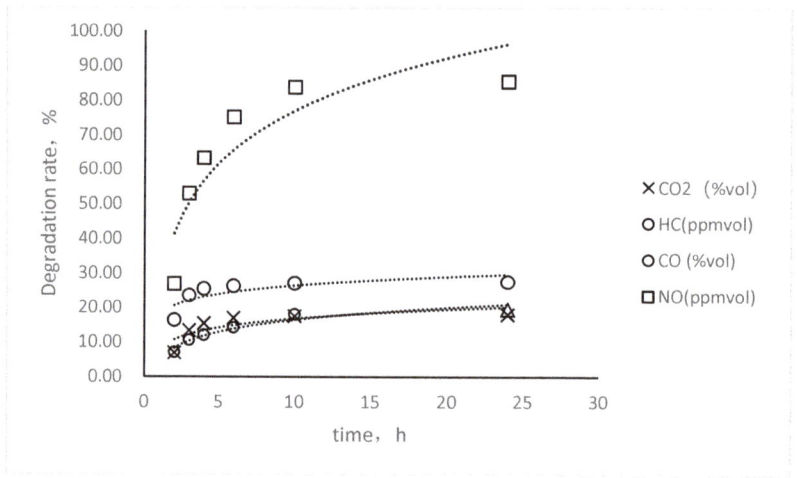

Figure 8. The effects of duration on degradation rate.

From the above experimental results, the overall trend is that the degradation rate of the exhaust gas increases with the extension of the reaction time, and the degradation rate curves of each gas generally agrees with the different amounts of variation of the nano-TiO$_2$. The degradation rate of CO and HC gas did not change much, but for the overall upward trend. From the trend line it can be seen that the three curves will eventually stay at a steady degradation rate, that is, CO$_2$ and CO will be close to 20%, HC close to 30%; the NO concentration will have a significant effect, more sustained and eventually will reach 85%. There is a more obvious linear correlation on the NO curve, accordingly, in the next study, the NO degradation rate is used as the recommended index.

The fact is that when the traffic flows in and out, greenhouse gases or harmful gases such as vehicle exhaust will be maintained at a specific concentration dynamically in a relatively confined space such as underground roads. The degradation effect of photocatalytic pavement will tend toward the best reduction rate as time goes by.

4.4. Different Light Intensity

The photocatalytic material nano-TiO$_2$ is added in the asphalt pavement. Ultraviolet irradiation is needed as an elementary condition for the degradation reaction. In this section, we continue to use the Section 3.2. light source arrangement; various numbers of xenon lamps were designed to simulate different light intensity to achieve different luminosity on the impact of exhaust degradation. The tests lasted 2 h. The degradation rates of the four kinds of harmful exhaust gases at different light intensities are shown in Figure 9.

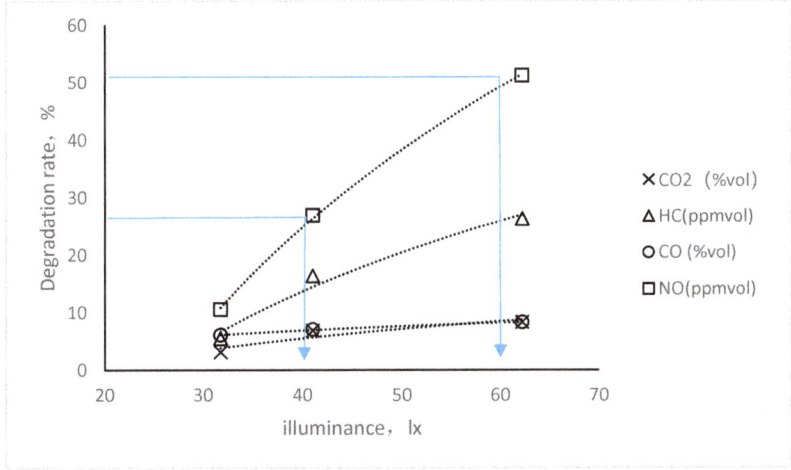

Figure 9. Fitting curves of degradation rates at different light intensities.

The statistical curves in Figure 9 suggest that illuminance and degradation rate have positive correlation, and the positive correlation tendency is very obvious on HC and NO. After 2 h, for Scheme C, the degradation rate of HC and NO reached 32.8% and 64.71%, respectively, which were considerable. Further, the proposed illumination index of lighting arrangement should meet some requirements (set NO degradation rate at 2 h as an indicator). To make the index more than 50% or 25%, the average illumination of road surface needs to be not less than 60 lx or 40 lx, respectively. The data show that the NO degradation rate of this method is higher than that described in other literature, under the basic similar and convertible conditions [7–9].

5. Conclusions

In this research program, test schemes were designed for the factors that may affect the performance of asphalt mixture on exhaust gas degradation, and the different influence factors were tested. According to the experimental test results and the statistical analysis findings, the following conclusions can be drawn:

1. The 5-nm particle size of TiO_2 is better than 10–15 nm on exhaust gas degradation, especially for HC and NO.

2. The experimental data showed that with an increase of the amount of photocatalytic material, the degradation of CO and CO_2 in the exhaust gas did not obviously increase, while the degradation effects of HC and NO were remarkable.

3. In the 4-h time extended degradation test, the experimental data show that the extended time has little effect on the degradation rate of CO_2 and CO, and the general trend is that the degradation of exhaust became significant with the extension of time.

4. 2 h NO degradation rate is set as an indicator. In order to make the index more than 50% or 25%, the average illumination of road surface needs to be not less than 60 lx or 40 lx.

Author Contributions: conceptualization, X.W.; methodology, M.H.; validation, M.H.; formal analysis, Ming Huang; resources X.W.; data curation, M.H.; writing—original draft preparation, M.H.; writing—review and editing, M.H.; visualization; supervision, X.W.; project administration, X.W.

Funding: 2015 and 2018 Technology Program of Shanghai Municipal Engineering Design Institute (Group) Co., Ltd. No. K2015K024 and K2018K081.

Acknowledgments: The authors would like to thank He Changxuan (SHMPI), Lv Weimin (Tongji University) and Dr. Zheng Xiaoguang (SMEDI) for helpful and constructive prophase studies.

Conflicts of Interest: The authors declare no conflict of interest.

References

1. Keyte, I.J.; Albinet, A.; Harrison, R.M. On-road traffic emissions of polycyclic aromatic hydrocarbons and their oxy-and nitro-derivative compounds measured in road tunnel environments. *Sci. Total Environ.* **2016**, *566*, 1131–1142. [CrossRef] [PubMed]
2. Moretti, L.; Cantisani, G.; Di Mascio, P. Management of road tunnels: Construction, maintenance and lighting costs. *Tunn. Undergr. Space Technol.* **2016**, *51*, 84–89. [CrossRef]
3. Chuan, L.; Xiaofeng, H.; Zijuan, L.; Lingyan, H. A Tunnel Test for $PM_{2.5}$ Emission Factors of Moto Vehicles in Shenzhen. *Environ. Sci. Technol.* **2012**, *35*, 150–153.
4. Cai, X. Application of Warm Mixed Asphalt Technology in Pavement of Urban Long Tunnel. *Urban Roads Bridges Flood Control* **2015**, *8*, 253–255.
5. Hurley, G.; Prowell, B. *Evaluation of Aspha-Min® Zeolite for Use in Warm Mix Asphalt*; NCAT Report 05-04; National Center for Asphalt Technology: Auburn, AL, USA, 2005; Available online: http://dsp2002.eng.auburn.edu/research/centers/ncat/files/reports/2005/rep05-04.pdf (accessed on 15 May 2019).
6. Tan, Y.; Li, L.; Wei, P.; Sun, Z. Application Performance Evaluation on Material of Automobile Exhaust Degradation in Asphalt Pavement. *China J. Highw. Transp.* **2010**, *23*, 21–27.
7. Wang, D.; Leng, Z.; Hüben, M.; Oeser, M.; Steinauer, B. Photocatalytic pavements with epoxy-bonded TiO_2-containing spreading material. *Constr. Build. Mater.* **2016**, *107*, 44–51. [CrossRef]
8. Toro, C.; Jobson, B.; Haselbach, L.; Shen, S.; Chung, S.; Chung, S. Photoactive roadways: Determination of CO, NO and VOC uptake coefficients and photolabile side product yields on TiO_2 treated asphalt and concrete. *Atmos. Environ.* **2016**, *139*, 37–45. [CrossRef]
9. Zhang, W. Experimental Studies on Automobile Exhaust Photocatalytic Degradationly Asphalt Pavement Material. Ph.D. Thesis, Chang'an University, Xi'an, China, 2014.
10. Strini, A.; Cassese, S.; Schiavi, L. Measurement of benzene, toluene, ethylbenzene and o-xylene gas phase photodegradation by titanium dioxide dispersed in cementitious materials using a mixed flow reactor. *Appl. Catal. B* **2005**, *61*, 90–97. [CrossRef]
11. Langridge, J.M.; Gustafsson, R.J.; Griffiths, P.T.; Cox, R.A.; Lambert, R.M.; Jones, R.L. Solar driven nitrous acid formation on building material surfaces containing titanium dioxide: A concern for air quality in urban areas? *Atmos. Environ.* **2009**, *43*, 5128–5131. [CrossRef]
12. Shen, S.; Burton, M.; Jobson, B.; Haselbach, L. Pervious concrete with titanium dioxide as a photocatalyst compound for a greener urban road environment. *Constr. Build. Mater.* **2012**, *35*, 874–883. [CrossRef]
13. Liu, W.; Wang, S.; Zhang, J.; Fan, J. Photocatalytic degradation of vehicle exhausts on asphalt pavement by TiO_2/rubber composite structure. *Constr. Build. Mater.* **2015**, *81*, 224–232. [CrossRef]
14. Luo, S.; Qian, Z.; Wang, H. Condition survey and analysis of epoxy asphalt concrete pavement on Second Nanjing Yangtze River Bridge: A ten-year review. *J. Southeast Univ.* **2011**, *27*, 417–422.
15. ASTM D1763-00(2013). Standard Specification for Epoxy Resins. Available online: https://www.astm.org/Standards/D1763.htm (accessed on 12 June 2019).
16. Huang, M.; Wen, X.; Wang, L. Influence of foaming effect, operation time and health preserving properties of foam epoxy asphalt mixtures. *Constr. Build. Mater.* **2017**, *151*, 931–938. [CrossRef]
17. JTJ F40-2004. *Technical Specifications for Construction of Highway Asphalt Pavements*; Communications Press: Beijing, China, 2004.
18. JTG E20-2011. *Standard Test Method of Bitumen and Bituminous Mixtures for Highway Engineering*; Communications Press: Beijing, China, 2011.
19. Hassan, M.M.; Dylla, H.; Asadi, S.; Mohammad, L.N.; Cooper, S. Laboratory Evaluation of Environmental Performance of Photocatalytic Titanium Dioxide Warm-Mix Asphalt Pavements. *J. Mater. Civ. Eng.* **2012**, *24*, 599–605. [CrossRef]
20. JTG/T D70/2-01-2014. *Guidelines for Design of Lighting of Highway Tunnels*; Communications Press: Beijing, China, 2014.

21. Wang, D.; Leng, Z.; Yu, H.; Hüben, M.; Kollmann, J.; Oeser, M. Durability of epoxy-bonded TiO$_2$-modified aggregate as a photocatalytic coating layer for asphalt pavement under vehicle tire polishing. *Wear* **2017**, *382*, 1–7. [CrossRef]
22. Wang, C.; Yan, K.; Han, X.; Shi, Z.; Li, B.; Feng, X.; Ping, N.; Wu, S. Physico-chemical Characteristic Analysis of PM$_{2.5}$ in the Highway Tunnel in the Plateau City of Kunming. *Environ. Sci.* **2017**, *38*, 4968–4975.

 © 2019 by the authors. Licensee MDPI, Basel, Switzerland. This article is an open access article distributed under the terms and conditions of the Creative Commons Attribution (CC BY) license (http://creativecommons.org/licenses/by/4.0/).

Article

Life Cycle Assessment for the Production Phase of Nano-Silica-Modified Asphalt Mixtures

Solomon Sackey [1], Dong-Eun Lee [2] and Byung-Soo Kim [1,*]

[1] Department of Civil Engineering, Kyungpook National University, 80 Daehakro, Daegu 41566, Korea; s.sackey123@knu.ac.kr
[2] Department of Architectural Engineering, Kyungpook National University, 80 Daehakro, Daegu 41566, Korea; dolee@knu.ac.kr
* Correspondence: bskim65@knu.ac.kr, Tel.: +82-10-6205-5348

Received: 8 March 2019; Accepted: 26 March 2019; Published: 29 March 2019

Featured Application: The application of LCA to NMAM has the potential to guide decision-makers on the selection of pavement modification additives to realize the benefits of using nanomaterials in pavements while avoiding potential environmental risks.

Abstract: To combat the rutting effect and other distresses in asphalt concrete pavement, certain modifiers and additives have been developed to modify the asphalt mixture to improve its performance. Although few additives exist, nanomaterials have recently attracted significant attention from the pavement industry. Several experimental studies have shown that the use of nanomaterials to modify asphalt binder results in an improved oxidative aging property, increased resistance to the rutting effect, and improves the rheological properties of the asphalt mixture. However, despite the numerous benefits of using nanomaterials in asphalt binders and materials, there are various uncertainties regarding the environmental impacts of nano-modified asphalt mixtures (NMAM). Therefore, this study assessed a Nano-Silica-Modified Asphalt Mixtures in terms of materials production emissions through the Life Cycle Assessment methodology (LCA), and the results were compared to a conventional asphalt mixture to understand the impact contribution of nano-silica in asphalt mixtures. To be able to compare the relative significance of each impact category, the normalized score for each impact category was calculated using the impact scores and the normalization factors. The results showed that NMAM had a global warming potential of 7.44563×10^3 kg CO_2-Eq per functional unit (FU) compared to 7.41900×10^3 kg CO_2-Eq per functional unit of the conventional asphalt mixture. The application of LCA to NMAM has the potential to guide decision-makers on the selection of pavement modification additives to realize the benefits of using nanomaterials in pavements while avoiding potential environmental risks.

Keywords: nanomaterials; life cycle assessment; nano-modified asphalt materials; environmental impact

1. Introduction

Asphalt is the most widely used pavement layer in the world. It consists of a binding material called bitumen and crushed or natural aggregates. The mixture of these materials forms asphalt mixtures. Demand for paved roads exceeded the supply of lake asphalts in the late 1800s and led to the use of petroleum asphalts [1]. Asphalt is often used as a shortened form of asphalt concrete which is the material of choice in the pavement sector. In the United Kingdom and the rest of Europe, the term 'bitumen' is used as a synonym for the term 'asphalt binder' while 'asphalt cement' is often used in the United States [2]. Asphalt cement or bitumen is used to bind the aggregates together to provide the required strength and stiffness to transfer vehicular loads. In addition to its strength and stiffness, asphalt pavements offer a damping ability due to the viscous-elastic nature of the bitumen [3].

Consequently, asphalt mixtures are qualified to provide optimal driving comfort as well as flexible maintenance actions. Asphalt pavements are designed to provide maximum performance throughout the design life. Bitumen (asphalt binder) performs two functions: Binding aggregates together and protecting the aggregates from distortions. However, unlike concrete pavements, asphalt pavements experience deformations over short periods of time. This, coupled with increased traffic loads and extreme weather conditions have resulted in asphalt pavement authorities seeking alternative solutions to improve the resistance of the road pavements to the adverse effects of mechanical and environmental loading [4].

Currently, several additives and modifiers produced commercially are used to modify the properties of the asphalt binder. Ref. [3] stated that additives and other modifiers are added in asphalt mixtures to lower mixing and compaction temperatures. This was found to improve adhesion and increase resistance against cracking and rutting. Regarding the viscosity of bitumen, Ref. [5] studied the effects of asphaltene on rheological properties of diluted Athabasca bitumen. Nanotechnology and nanomaterials have recently attracted significant attention from the pavement industry. Nanomaterial application is considered to have the potential to improve asphalt binder properties. As mentioned by Ref. [6], the application of nanomaterials as asphalt modifiers is growing rapidly in popularity due to its unique characteristics that significantly improve the performance of asphalt binder. It has been shown in several studies that the addition of nano-silica in asphalt mixtures improve the oxidative aging property, increases resistance to the rutting effect, improves the rheological properties of asphalt mixture and decreases the interaction between asphalt molecules [7–10]. In addition, Ref. [11] investigated and found that increasing nano-silica content in asphalt mixtures decreases the ductility and temperature sensitivity of the asphalt mixture.

It is becoming increasingly important to explore the full benefits of additives and modifiers on the long-term performance of asphalt pavements. With sustainability in mind, and also embracing the global effort to reduce the environmental impacts associated with these newly perforated materials, being able to make decisions and judge the benefits and environmental friendliness linked to the long-term pavement performance has become important. Consequently, having a life-cycle assessment (LCA) tools available to assess modified-asphalt materials on a life-cycle basis becomes necessary. Due to the concern of global warming and resource depletion, LCAs for different materials and products and systems have gained significant popularity with researchers. LCA studies can help to determine and minimize the energy consumption, use of resources, and emissions to the environment by providing a superior understanding of the systems [3]. LCA studies can help to consider different alternatives if the environmental performance of a particular material or product is not favorable. There have been several studies that attempt to assess the environmental impact of asphalt materials and some studies have also been conducted on asphalt binders modified with additives [12–15]. However, to the author's knowledge, no studies that assess the complete LCA for the production phase of nano-silica-modified asphalt mixtures have previously been published. A new material being used as a modifier, there are uncertainties regarding the environmental impacts associated with nanomaterials. Therefore, it is of paramount importance to investigate the extent to which the use of nano-silica-modified asphalt mixtures for asphalt concrete pavement is beneficial from an environmental perspective.

This study presents the assessment of a Nano-Silica-Modified Asphalt Mixtures in terms of materials production emissions through LCA methodology. The environmental impacts of a conventional asphalt mixture were assessed so that a comparison could be made to understand the impact contribution of nano-silica in the asphalt mixture. In addition, to be able to compare the relative significance of each impact category, the normalized score was computed for each impact category using impact scores and normalization factors. The application of LCA to NMAM has the potential to guide decision-makers on the selection of pavement modification additives to realize the full benefits of the use of nanomaterials in pavements while avoiding potential environmental risks.

2. Literature Review and Definitions

2.1. Life Cycle Assessment

LCA is described by Ref. [16] as a tool for systematically analyzing the environmental performance of products or processes over their entire life cycle, which includes raw material extraction, manufacturing, use, end-of-life disposal, and recycling. LCA is described as a 'cradle to grave' method for the evaluation of environmental impacts [17]. In a similar description, Ref. [18] defines LCA as a methodology that quantifies the environmental impacts of a process or a product. In their study, Ref. [19] stated that most of the environmental impacts do not occur in the use, maintenance, and repair of the product but during the manufacturing, transportation, and disposal stages. Ref. [20] claimed that it would be premature to make any claims on the environmental benefits of a particular product or manufacturing process without first considering its consequences in a life cycle context. LCA methodology includes the establishment of an inventory of all types of emissions and waste products [21,22]. LCA studies are conducted in accordance with the specification and standards of the International Organization for Standardization (ISO). The four major components of an LCA study according to Ref. [23] are illustrated in Figure 1. The inventory analysis part is made up of material extraction phase, manufacturing or production phase, use or operational phase and disposal phase. However, it is quite difficult to effectively assess the environmental impact of a product during its in-service life. Therefore, the analysis of this study does not include the operational phase and/or the disposal phase of the inventory analysis.

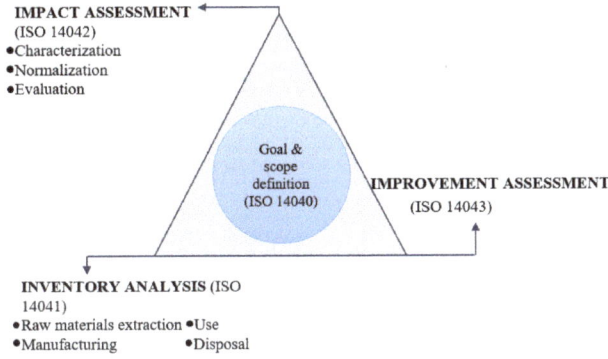

Figure 1. Structure of LCA study.

2.2. Nanomaterials and their Application as a Modifier in Asphalt Mixtures

Nanotechnology is an emerging technology and is regarded as a key enabling technology due to its numerous associated benefits to many areas of society. Nanotechnology is defined as the use of very small particles of materials (either by themselves or by their manipulation) to create new large materials [24]. The author added that nanotechnology is not a new science or technology, but an extension of the science and technology that has been in development for many years, and is used to examine nature at an ever-smaller scale. Ref. [25] defines nanomaterials as those physical substances with at least one dimension between 1 and 150 nm (1 nm = 10^{-9} m). With reference to the European Commission's recommended definition of nanomaterials, Ref. [26] defines nanomaterial as a "natural, manufactured material containing particles, in an unbound state or as an aggregate or as an agglomerate and where, for 50% or more of the particles in the number size distribution, one or more external dimensions is in the size range 1–100 nm". The application of nanomaterials in the field of construction is growing rapidly. Ref. [27] mentioned that nanotechnology is a rapidly expanding area of research where novel properties of materials manufactured at the nanoscale can be utilized for the benefits of constructing infrastructure. Although some nanomaterials are already being used in the concrete industry, their application as a modifier in asphalt binder has attracted more interest recently. Several experimental studies have been conducted to determine the effect of nanomaterials, especially nano-silica on the properties of asphalt mixtures. Nano-silica materials are used as additives which are applied in small percentages by weight of the asphalt binder to improve the rheological and other properties of asphalt mixtures. Ref. [7] investigated the characteristics of asphalt binder and mixture containing nano-silica and found that the addition of nano-silica has a positive influence on different properties of the asphalt binder and mixture. Ref. [28] also studied the effect of nano-silica and rock asphalt on rheological properties of modified bitumen. In their study, Ref. [29] found that the inclusion of nano-silica reduces the rutting susceptibility of nano-modified asphalt mixtures. Ref. [30] studied laboratory evaluation of composed modified binder and mixture containing nano-silica/rock asphalt/SBS. In a similar experimental study, Ref. [31] found that increasing the percentage of nano-silica increases the Brookfield Rotational Viscosity (RV). Ref. [32] worked on the application of nano-silica to improve asphalt mixture self-healing. In another study, Ref. [33] investigated the effect of nano-silica on thermal sensitivity of hot-mix asphalt. Nano-silica increases the strength or durability of asphalt mixture [34,35]. Refs. [36–40] also made similar studies on the effect of nano-silica on asphalt binder and mixtures. Table 1 summarizes the review of previous studies on the characterization of asphalt binder modified with nano-silica. Regarding the cost of using nanomaterials, Ref. [41] provides the prices for almost all nanomaterials based on the quantity required. For example: precipitated calcium carbonate Nanopowder, 50 nm (100 g = $45, 1 kg = $85); nano-silica nanopowder, 60–70 nm (100 g = $55, 1 kg = $155); titanium oxide Nanopowder, 20 nm (100 g = $165, 1 kg = $468); zinc oxide Nanopowder, 80–200 nm (100 g = $58, 1 kg = $168). While some nanomaterials may seem costly, others may be cheap. However, on a large scale, an extensive economic analysis is required to determine the optimum cost for each nanomaterial based on the quantity required.

Table 1. Review of previous studies on modification of asphalt binder with nano-silica.

Author	Type of Nanomaterial	Effect on Asphalt Binder and Mixtures
[32]	Nano-silica	Improves the self-healing of HMA
[10]	Nano-silica	Improves marshal stability, resilient modulus, and fatigue life
[29]	Nano-silica	Enhances antiaging property and rutting and fatigue cracking performance
[30]	Nano-silica	Improves temperature stability, decreases temperature cracking resistance and reduces susceptibility to moisture damage
[28]	Nano-silica	Enhances the complex shear modulus and improves the anti-rutting performance of asphalt mixture
[34]	Nanosilica	Reduces the susceptibility to moisture damage and increases the strength of asphalt mixes
[35]	Nano-silica	Improves the performance and durability of asphalt mixtures
[36]	Nano-silica	Improve rutting and fatigue performance of asphalt binder
[37]	Nano-silica	Decreases the interaction between asphalt molecules and increases free volumes in the configuration
[38]	Nano-silica	Decreases the consistency, rate of water absorption and porosity of the roller compacted concrete pavement
[39]	Nano-silica	Improves the rheological characteristics, toughness, and viscosity of bitumen
[40]	Nano-silica	Reduces the creep strain deformation and increases the dynamic shear modulus

3. Methodology

LCA methodology was used (as standardized by the ISO in 2006) to assess the environmental impact of nano-silica-modified asphalt mixtures. There are numerous nanomaterials whose effect on asphalt binder and mixtures have previously been evaluated. However, based on the extensive literature review, the common nanomaterials which have been experimentally shown to have a greater impact on asphalt concrete performance include nanoclay and nano-silica. Consequently, nano-silica was used in this study. However, any other nanomaterial (especially nanoclay) which uses a similar production process could give similar results when modified with asphalt binder and materials. Also, the analysis of this study focused on only material extraction and production phases and does not include the operational or the disposal phase. The inclusion of the operational phase in the LCA analysis could change the inference about the conformity of nano-silica-modified asphalt mixtures.

The structure of LCA studies adopted includes goal and scope definition, inventory analysis, impact assessment, and improvement assessment or interpretation stages.

3.1. Goal and Scope Definition

3.1.1. Goal and System Boundaries

The goal of this study is to assess the potential life-cycle environmental impacts resulting from modifying asphalt materials with nanomaterial (i.e., the environmental impacts of nano-silica-modified asphalt mixtures). Additionally, a comparison is made with the environmental impacts of unmodified asphalt mixture to provide a better understanding of the impact contribution of nanomaterials in asphalt materials to allow for informed decisions to be made. In other words, the extent to which the use of nano-silica-modified asphalt mixtures for asphalt concrete pavement is beneficial from the environmental perspective is evaluated.

Two alternative case scenarios were examined. In CASE 1A, the environmental impact of nano-modified asphalt material was assessed. The use of nanomaterial (nano-silica), asphalt materials, and the production processes of asphalt mixtures were considered. Modification of bitumen with nanomaterial is depicted in Figure 2.

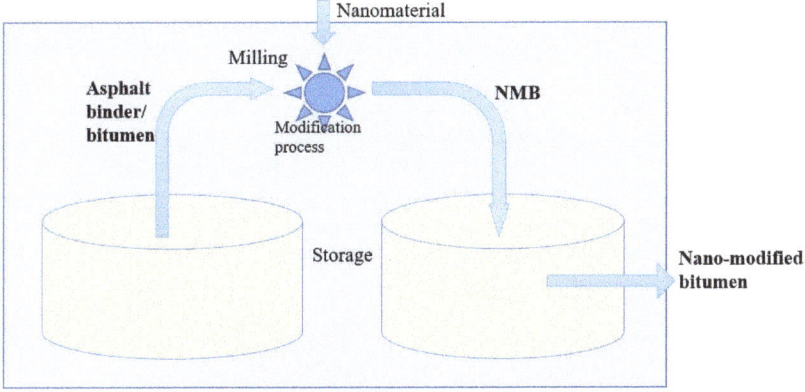

Figure 2. Nano-modified bitumen (NMB) production.

In CASE 2A, the environmental impact of asphalt material production, excluding nanomaterial (conventional asphalt mixture), was assessed. The system boundaries which defines the unit process considered in the LCA studies [4] were limited to cover the following life cycle stages in this study: (1) raw materials extraction; (2) transportation of raw materials for a unit product manufacturing; (3) modification and production of asphalt materials in the plant. Transportation of asphalt materials to the field, use, and the end-of-life were not included. The life cycle stages and key processes of nano-modified asphalt production in the plant are shown in Figure 3. The flow emissions and resource consumption (such as electricity and natural gas) for heating and the production processes were also included in the system boundaries.

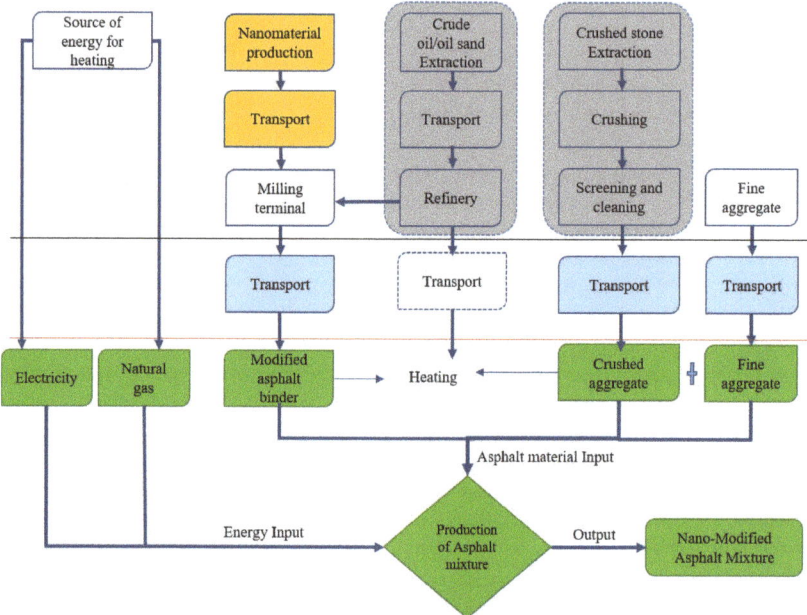

Figure 3. Key processes of nano-modified asphalt materials.

The results of this study will help practitioners in the asphalt concrete pavement industry to make informed decisions by considering the numerous benefits of nanomaterials (nano-silica) and the environmental impacts resulting from modifying asphalt mixtures with the nano-silica material.

3.1.2. Functional Unit (FU)

The FU is the heart of any LCA studies. The FU is a quantified performance of a product system for use as a reference unit in an LCA study [21] (referring to the Malaysian standards handbook on environmental management). A fixed value must be created and the output results of the environmental impacts of the impact categories depend on this selected FU. In this study, a FU of 1000 kg production of nano-silica-modified asphalt mixtures was assumed.

3.2. *Life Cycle Inventory (LCI)*

Material Extraction and Production Processes

The life cycle inventory stage is the stage of actual data collection and the modeling of the system product. For the data on material extraction, processing, and production, an openLCA database was used. OpenLCA is an open source LCA tool from GreenDeLTa located in Berlin, Germany. The software uses an Eco-invent 2.2 database and other proprietary databases and produces equally good results compared to other proprietary LCA tools such as SimaPro, Gabi, etc. The software allows the user to import any external database into its platform and it can be used to model any product. For the production process of nano-silica, silica gel and precipitated silica type, the process outlined by Ref. [42] was followed. A 1000 kg production of bitumen and aggregates was assumed. For the input amount of

1kg nano-silica production, the data provided by Ref. [43] were referred to. Additives are often applied in small percentages (1–10%) by weight of asphalt binder. This study used 3% of nano-silica for asphalt binder modification. Therefore, the input amount of 30 kg nano-silica was required to modify the bitumen. Data from Refs. [13,14] were used regarding the energy consumption data per kg of material required for bitumen production, aggregates, and the mixing of asphalt materials at the plant. In other studies, such as the one reported by Ref. [3], the modification of asphalt binder with additives results in an increase in fuel consumption by approximately 15%. Therefore, it was assumed that an increase of 15% in energy for bitumen production was required to modify the bitumen. Hence, to account for the asphalt binder modification with nano-silica in the analysis, an additional 15% increase in energy (fuel) was added to the 0.51 MJ energy for bitumen production. Transportation of nano-silica material to the milling terminal for modification was assumed as 100 km, while the total transportation for bitumen and to the asphalt plant was also assumed to be 100 km and that for aggregates was assumed as 5 km to the mixing plant. In Table 2, the material and energy requirements for the production of nanomaterial (nano-silica) and asphalt materials is summarized.

Table 2. Materials and energy requirements for 1 kg unit production of nano-silica and asphalt materials.

Input	Flow Amount
Nanomaterial (nano-silica)	
Sodium silicate	0.66 kg
Sulfuric acid	3.9 kg
Heat (Natural gas)	15–24 MJ
Water	40 kg
Asphalt material	
Total Energy for bitumen production	0.51 MJ
Energy for aggregates production	0.0354 MJ
Energy for asphalt materials production	0.349 MJ

3.3. Life Cycle Impact Assessment (LCIA)

The life cycle impact assessment (LCIA) stage involves analyzing the data to evaluate the contribution to each impact category. LCIA consists of characterization, normalization, evaluation, and weighting depending on the LCIA used. In this study, the Tool for the Reduction and Assessment of Chemical and other Environmental Impacts (TRACI) method version 2.1 (provided in openLCA) was used to calculate the impact category indicator scores TRACI is a software from the US Environmental Protection Agency (EPA), Durham, NC, USA. TRACI uses Equation (1) [44] to determine the impact score for each individual environmental impact category.

$$I^i = \sum_{xm} CF^i_{xm} \times M_{xm} \tag{1}$$

where: I^i is the potential impact of all substances (x) for a specific category of concern (i), CF^i_{xm} is the characterization factor for substance (x) emitted to media (m) for each impact category (i), and M_{xm} is the mass of the substance emitted to media (m). OpenLCA version 1.7.4 was then used for modeling the processes in this study.

Finally, to be able to compare the relative significance of each impact category, the normalized score for each impact category was calculated. Normalization is the ratio of the impact score in each category and the estimated impacts from a reference (often called normalization factors). These factors represent the impact produced by an average person in a reference place per year. Equation (2) was used for the computation:

$$NS_i = \frac{EnvI_i}{NF_i} \tag{2}$$

where NS_i is the normalized score of impact category i, $EnvI_i$ is the environmental impact result of impact category i, and NF_i is the normalization factor of impact category i. Regarding the normalization factors, US 2008 reference data was used, the impact per person-year updated in the research by Ref. [40]. Table 3 provides the details of the normalization factors. The units of four categories: ecotoxicity, carcinogenic, non-carcinogenic, and acidification are different from the reference units first converted to the reference units before computing the normalized score.

Table 3. Normalization factors for impact categories based on inventories from the US (2008) and US-Canada [45].

Impact Category	Normalization Factors and Reference Year			
	US 2008		US-CA 2005/2008	
	Impact per Year	Impact per Person Year	Impact per Year	Impact per Person Year
Ecotoxicity-metals (CTUe)	3.30×10^{12}	1.10×10^4	3.70×10^{12}	1.10×10^4
Ecotoxicity-non-metals (CTUe)	2.30×10^{10}	7.60×10^1	2.50×10^{10}	7.40×10^1
Carcinogens-metals (CTUcanc.)	1.40×10^4	4.50×10^{-5}	1.50×10^4	4.30×10^{-5}
Non-carcinogens-metals (CTUcanc.)	3.10×10^5	1.00×10^{-3}	3.40×10^5	1.00×10^{-3}
Global warming (kg CO_2 eq)	7.40×10^{12}	2.40×10^4	8.00×10^{12}	2.40×10^4
Ozone depletion (kg CFC-11 eq)	4.90×10^7	1.60×10^{-1}	4.90×10^7	1.50×10^{-1}
Acidification (kg SO_2 eq)	2.80×10^{10}	9.10×10^1	3.20×10^{10}	9.50×10^1
Eutrophication (kg N eq)	6.60×10^9	2.20×10^1	7.00×10^9	2.10×10^1
Photochemical ozone formation (kg O_3 eq)	4.20×10^{11}	1.40×10^3	4.90×10^{11}	1.50×10^3
Respiratory effects (kg PM2.5 eq)	7.40×10^9	2.40×10^1	1.00×10^{10}	3.00×10^1

Acidification potential = 1.98×10^{-2} SO_2/kg substance (multiplied its impact result by this value).

Ecotoxicity potential for rural air = 0.064 CTU eco/kg substance (multiplied its impact result by this value).

Human health cancer potential for rural air = 1.2×10^{-7} CTU canc/kg substance (multiplied its impact result by this value).

Human health non-cancer potential for rural air = 3.0×10^{-8} CTU canc/kg substance (multiplied its impact result by this value).

4. Results and Discussion

4.1. CASE 1A: Impact Assessment of Nano-Silica-Modified Asphalt Mixtures Analysis

OpenLCA version 1.7.4 was used to model and analyzed the environmental impacts of nano-modified asphalt materials and the analysis results are shown in Table 4.

Table 4. LCIA results of nano-silica-modified asphalt mixtures per FU.

Impact Category	Reference Unit	Impact Result
Environmental impact ǀ global warming	kg CO_2-Eq	7.44563×10^3
Human health ǀ respiratory effects, average	kg PM2.5-Eq	8.86935×10^2
Environmental impact ǀ ozone depletion	kg CFC-11-Eq	3.71600×10^{-2}
Environmental impact ǀ eutrophication	kg N-Eq	1.49156×10^1
Human health ǀ carcinogenic	kg benzene-Eq	2.18467×10^3
Environmental impact ǀ photochemical oxidation	kg NOx-Eq	3.03420×10^1
Human health ǀ non-carcinogenics	kg toluene-Eq	6.07040×10^6
Environmental impact ǀ ecotoxicity	kg 2,4-D-Eq	1.08917×10^4
Environmental impact ǀ acidification	moles of H+-Eq	1.87879×10^5

Increase in the production of the raw materials and/or the FU results in an increase in fuel and energy usage and will cause a significant increase in the impact scores in each category. The environmental performance of 7.44563×10^3 kg CO_2-Eq/FU global warming of nano-silica-modified asphalt mixture is better than the results of (Butt et al.) who found the modification of asphalt materials with a polymer to be 44.9×10^3 kg CO_2-Eq per FU of 1 km by 3.5 km wide asphalt pavement.

4.2. CASE 2A: Impact Assessment of Unmodified (Conventional) Asphalt Mixture Analysis

The analysis of unmodified (conventional) asphalt materials was needed to better understand the environmental implication of modifying conventional asphalt with nanomaterials. The results of the analysis are shown in Table 5.

Table 5. LCIA results of unmodified (conventional) asphalt materials per FU.

Impact Category	Reference Unit	Impact Result
Environmental impact ǀ global warming	kg CO_2-Eq	7.41900×10^3
Human health ǀ respiratory effects, average	kg PM2.5-Eq	8.79600×10^2
Environmental impact ǀ ozone depletion	kg CFC-11-Eq	3.68100×10^{-2}
Environmental impact ǀ eutrophication	kg N-Eq	1.47700×10^1
Human health ǀ carcinogenic	kg benzene-Eq	2.16300×10^3
Environmental impact ǀ photochemical oxidation	kg NOx-Eq	3.01270×10^1
Human health ǀ non-carcinogenics	kg toluene-Eq	6.01233×10^6
Environmental impact ǀ ecotoxicity	kg 2,4-D-Eq	1.08133×10^4
Environmental impact ǀ acidification	moles of H+-Eq	1.86282×10^5

Any increase in the production of raw materials or a change in the FU will result in an increase in the impact scores in each category and vice versa.

The modification of asphalt materials with nanomaterials results in an increase in environmental impacts, which is clear when comparing the results in Table 5 with that in Table 4. Across all impact categories, there is an increase in the impact scores. This fact is reinforced by Ref. [4] when the authors found that using Ethylene-Vinyl-Acetate (EVA) polymer as a modifier agent leads to a deterioration of the life cycle profile of the pavement compared to unmodified asphalt binder. However, the deterioration of the life cycle environmental profile with nano-modified asphalt materials is insignificant. Specifically, there was only a 0.4% increase in global warming, 0.8% increase in respiratory effects, 0.009% increase in ozone depletion, 0.98% increase in eutrophication, 1.0% increase in human health carcinogenic, 0.7% increase in photochemical oxidation, 0.96% increase in human health non-carcinogenic, 0.72% increase in ecotoxicity, and 0.85% increase in acidification. This means the modification of asphalt materials with nanomaterials (nano-silica) causes more impacts in human health carcinogenic than other impact categories. Apart from ozone depletion, the modification of asphalt materials with nano-silica contributes fewer impacts in global warming per 3% by weight of asphalt binder production of nano-silica.

4.3. Computation of Normalised Score

Table 6 and Figure 4 show the normalized score in each impact category of nano-modified asphalt materials. According to Ref. [46], by inspection, large values of normalized scores as compared to the total are classified as worse performing impact categories, while those with small normalized scores of approximately less than 2% of the total are classified as better performing impact categories. Table 6 shows that nano-modified asphalt materials only perform significantly better in four impact categories: photochemical oxidation (0.0217 pts/FU), ecotoxicity (0.0634 pts/FU), ozone depletion (0.2323 pts/FU), and global warming (0.3102 pts/FU).

However, to fully understand when and how an impact category is classified as either better or worse performing, a logarithmic scale criterion was used. This was especially useful in situations where there existed large variation in the normalization scores. It is argued by Ref. [47] that dimensionless data is more appropriately plotted on an arithmetic scale to clearly understand where the data points lie (better or worse trend). On a logarithmic scale, the center of gravity (where the eye is drawn) lies at the geometric mean, where the line starts at 1 and not 0. Hence, applying the logarithmic scale plot (see Figure 4), all the impact categories below the 1pts line are referred to as ZONE 1 (better performance zone). Hence, it can be said NMAM performs better in five categories: global warming (0.3102 pts/FU), ozone depletion (0.2323 pts/FU), eutrophication (0.6779 pts/FU), photochemical oxidation (0.0217 pts/FU), and ecotoxicity (0.0634 pts/FU).

Table 6. Normalized score per FU of the impact categories for NMAM.

Impact Category	Normalized Score (points, pts)
Environmental impact	global warming	0.3102
Human health	respiratory effects, average	36.9556
Environmental impact	ozone depletion	0.2323
Environmental impact	eutrophication	0.6779
Human health	carcinogenic	5.8258
Environmental impact	photochemical oxidation	0.0217
Human health	non-carcinogenic	182
Environmental impact	ecotoxicity	0.0634
Environmental impact	acidification	41.0101

Small values are better.

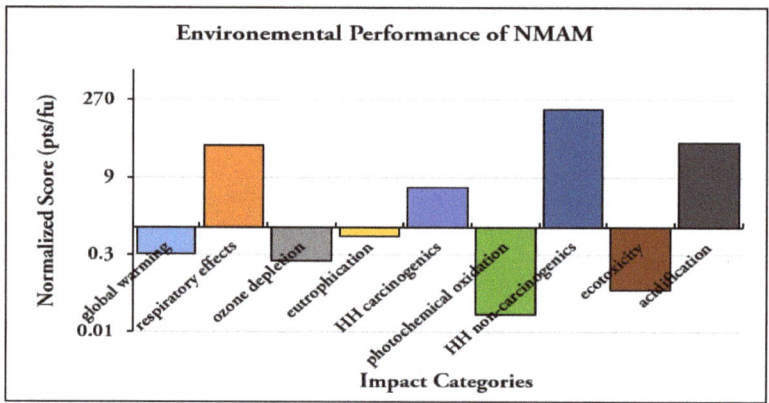

Figure 4. Environmental performance (normalized score) of NMAM.

All the impact categories above the 1 pts line are referred to as ZONE 2 (worse performance zone). NMAM performs worse in this zone in 4 categories: respiratory effects (36.9556 pts/FU), human health carcinogenic (5.8258 pts/FU), human health non-carcinogenic (182 pts/FU), and acidification (41.0101 pts/FU).

The worst performance in acidification, which is the increase in hydrogen ions (H+) concentration within the environment as a result of the presence of acids, can be attributed to the sulfuric acid used in the production of nano-silica and the cause of sulphur dioxide and nitrogen oxides released during transportation of the materials and including asphalt materials. As mentioned previously, the modification of asphalt materials with nanomaterial causes only 0.4% per unit increase in global warming. This is because carbon dioxide (the main cause of global warming) is released during the production of bitumen, aggregates, asphalt mixing, and also during transportation. In short, the fact that the modification of asphalt materials with nanomaterial causes just less than or equal to 1% increase in impact score across all impact categories suggests that modifying asphalt materials with nanomaterials does not cause an unreasonable risk to the environment. However, the results of this study using nano-silica does not conclude that all other nanomaterials may have very low impact. The impact on the environment and the combined impact when modified with asphalt materials depend on the production process of the nanomaterial. Therefore, it is expected that some nanomaterials may have a more negative environmental impact.

5. Conclusions

LCA is a tool that helps to assess the environmental impacts of materials and products so that decisions can be made not just on the benefits of using these materials but also considering their environmental contributions (especially to climate change and human health). This study assessed a Nano-Silica-Modified Asphalt Mixtures in terms of materials production emissions through the Life Cycle Assessment methodology (LCA), and the results were compared to a conventional asphalt mixture to understand the impact contribution of nano-silica in asphalt mixtures. The results showed that NMAM had a global warming potential of 7.44563×10^3 kg CO_2-Eq per FU as compared to 7.41900×10^3 kg CO_2-Eq per FU of unmodified asphalt mixture. The study also computed the normalized score for each impact category and the results showed NMAM performs better in five categories: global warming (0.3102 pts/FU), ozone depletion (0.2323 pts/FU), eutrophication (0.6779 pts/FU), photochemical oxidation (0.0217 pts/FU), and ecotoxicity (0.0634 pts/FU). NMAM performs worse in four categories: respiratory effects (36.9556 pts/FU), human health carcinogenic (5.8258 pts/FU), human health non-carcinogenic (182 pts/FU), and acidification (41.0101 pts/FU). The modification of asphalt materials with nano-silica causes less than or equal to 1% per unit increase in impact score across all impact categories. The application of LCA to NMAM has the potential to guide decision-makers on the selection of pavement modification additives to realize the benefits of nanomaterials in the pavement while avoiding potential environmental risks. Additionally, this study has shown that even though the modification of asphalt mixtures with nano-silica results in an increase in fuel consumption, it does not cause an unreasonable risk to the environment nor does its application as a modifier results in significant deterioration of the life cycle environmental profile. However, future research is required by considering the analysis of the whole life cycle for nano-modified asphalt materials using different nanomaterials as a modifier to confirm that nanomaterials are sustainable materials.

Author Contributions: Conceptualization, S.S.; Methodology, S.S.; Software, S.S.; Validation, S.S.; Formal Analysis, S.S.; Writing–Original Draft Preparation, S.S., D.E.L., B.S.K.; Writing–Review & Editing, B.S.K.; Funding Acquisition, D.E.L., B.S.K.

Funding: This work was supported by the National Research Foundation (NRF) of Korea grant funded by the Korea government (MSIT) (No.NRF-2018R1A5A1025137).

Conflicts of Interest: The authors declare no conflict of interest.

References

1. Blow, M.D. Asphalt Binder Basics Specifications, History and Future. In Proceedings of the North Dakota Asphalt Conference, Bismarck, ND, USA, 5 April 2016.
2. Hihara, L.H.; Adler, R.P.I.; Latanision, R.M. (Eds.) *Environmental Degradation of Advanced and Traditional Engineering Materials*; CRC Press: Boca Raton, FL, USA, 2013.
3. Butt, A.A.; Birgisson, B.; Kringos, N. Considering the benefits of asphalt modification using a new technical life cycle assessment framework. *J. Civ. Eng. Manag.* **2016**, *22*, 597–607. [CrossRef]
4. Santos, J.; Cerezo, V.; Soudani, K.; Bressi, S. A Comparative Life Cycle Assessment of Hot Mixes Asphalt Containing Bituminous Binder Modified with Waste and Virgin Polymers. *Procedia CIRP* **2018**, *69*, 194–199. [CrossRef]
5. Mozaffari, S.; Tchoukov, P.; Atias, J.; Czarnecki, J.; Nazemifard, N. Effect of asphaltene aggregation on rheological properties of diluted Athabasca bitumen. *Energy Fuels* **2015**, *29*, 5595–5599. [CrossRef]
6. Jamshidi, A.; Hasan, M.R.M.; Yao, H.; You, Z.; Hamza, M.O. Characterization of the rate of change of rheological properties of nano-modified asphalt. *Constr. Build. Mater.* **2015**, *98*, 437–446. [CrossRef]
7. Enieb, M.; Diab, A. Characteristics of asphalt binder and mixture containing nano-silica. *Int. J. Pavement Res. Technol.* **2017**, *10*, 148–157. [CrossRef]
8. Zhu, X.; Du, Z.; Ling, H.; Chen, L.; Wang, Y. Effect of filler on thermodynamic and mechanical behaviour of asphalt mastic: a MD simulation study. *Int. J. Pavement Eng.* **2018**, *24*, 1–15. [CrossRef]

9. Sadeghpour Galooyak, S.; Palassi, M.; Goli, A.; Zanjirani Farahani, H. Performance evaluation of nano-silica modified bitumen. *Int. J. Transp. Eng.* **2015**, *3*, 55–66.
10. Hasaninia, M.; Haddadi, F. The characteristics of hot mixed asphalt modified by nanosilica. *Pet. Sci. Technol.* **2017**, *35*, 351–359. [CrossRef]
11. Taherkhani, H.; Afroozi, S. The properties of nanosilica-modified asphalt cement. *Pet. Sci. Technol.* **2016**, *34*, 1381–1386. [CrossRef]
12. Samieadel, A.; Schimmel, K.; Fini, E.H. Comparative life cycle assessment (LCA) of bio-modified binder and conventional asphalt binder. *Clean Technol. Environ. Policy* **2018**, *20*, 191–200. [CrossRef]
13. Eurobitume. Life cycle inventory: Bitumen, European Bitumen Association. Available online: https://www.eurobitume.eu/fileadmin/pdf-downloads/LCI%20Report-Website-2ndEdition-20120726.pdf (accessed on 12 December 2018).
14. Mukherjee, A. *Life Cycle Assessment of Asphalt Mixtures in Support of an Environmental Product Declaration*; National Asphalt Pavement Institute: Lanham, MD, USA, 2016.
15. Ventura, A.; Monéron, P.; Jullien, A. Environmental impact of a binding course pavement section, with asphalt recycled at varying rates: use of life cycle methodology. *Road Mater. Pavement Des.* **2008**, *9*, 319–338. [CrossRef]
16. Sackey, S.; Kim, B.S. Environmental and economic performance of asphalt shingle and clay tile roofing sheets using life cycle assessment approach and topsis. *J. Constr. Eng. Manag.* **2018**, *144*, 04018104. [CrossRef]
17. Cabeza, L.F.; Rincon, L.; Vilarino, V.; Perez, G.; Castell, A. Life cycle assessment (LCA) and life cycle energy analysis (LCEA) of buildings and the building sector: A review. *Renew. Sustain. Energy Rev.* **2014**, *29*, 394–416. [CrossRef]
18. Lunardi, M.; Alvarez-Gaitan, J.; Bilbao, J.; Corkish, R. Comparative Life Cycle Assessment of End-of-Life Silicon Solar Photovoltaic Modules. *Appl. Sci.* **2018**, *8*, 1396. [CrossRef]
19. Joshi, S. Product environmental life-cycle assessment using input-output techniques. *J. Ind. Ecol.* **1999**, *3*, 95–120. [CrossRef]
20. Dhingra, R.; Kress, R.; Upreti, G. Does lean mean green? *J. Clean. Prod.* **2014**, *85*, 1–7. [CrossRef]
21. Sharaai, A.H.; Mahmood, N.Z.; Sulaiman, A.H. Life cycle impact assessment (LCIA) using TRACI methodology: An analysis of potential impact on potable water production. *Aust. J. Basic Appl. Sci.* **2010**, *4*, 4313–4322.
22. Park, W.J.; Kim, T.; Roh, S.; Kim, R. Analysis of Life Cycle Environmental Impact of Recycled Aggregate. *Appl. Sci.* **2019**, *9*, 1021. [CrossRef]
23. International Organization for Standardization. *Environmental Management: Life Cycle Assessment; Principles and Framework*; ISO: Geneva, Switzerland, 2006.
24. Mann, S. Nanotechnology and Construction; Nanoforum Report. 2006. Available online: https://nanotech.law.asu.edu/Documents/2009/10/Nanotech%20and%20Construction%20Nanoforum%20report_259_9089.pdf (accessed on 12 December 2018).
25. Olar, R. Nanomaterials and nanotechnologies for civil engineering. *Buletinul Institutului Politehnic din Iasi. Sectia Constructii, Arhitectura* **2011**, *57*, 109.
26. Lazarevic, D.; Finnveden, G. *Life Cycle Aspects of Nanomaterials*; KTH- Royal Institute of Technology: Stockholm, Sweden, 2013.
27. Singh, K.P. Nanotechnology and Nano-Materials: An Advance Developing Approach for Indian Industries. *IJARESM* **2018**, *6*, 4.
28. Shi, X.; Cai, L.; Xu, W.; Fan, J.; Wang, X. Effects of nano-silica and rock asphalt on rheological properties of modified bitumen. *Constr. Build. Mater.* **2018**, *161*, 705–714. [CrossRef]
29. Yao, H.; You, Z.; Li, L.; Lee, C.H.; Wynyard, D.; Yap, Y.K.; Shi, X.; Goh, S.W. Rheological properties and chemical bonding of asphalt modified with nanosilica. *J. Mater. Civ. Eng.* **2012**, *25*, 1619–1630. [CrossRef]
30. Cai, L.; Shi, X.; Xue, J. Laboratory evaluation of composed modified asphalt binder and mixture containing nano-silica/rock asphalt/SBS. *Constr. Build. Mater.* **2018**, *172*, 204–211. [CrossRef]
31. Ezzat, H.; El-Badawy, S.; Gabr, A.; Zaki, E.S.I.; Breakah, T. Evaluation of asphalt binders modified with nanoclay and nanosilica. *Procedia Eng.* **2016**, *143*, 1260–1267. [CrossRef]
32. Ganjei, M.A.; Aflaki, E. Application of nano-silica and styrene-butadiene-styrene to improve asphalt mixture self-healing. *Int. J. Pavement Eng.* **2019**, *20*, 89–99. [CrossRef]

33. Firouzinia, M.; Shafabakhsh, G. Investigation of the effect of nano-silica on thermal sensitivity of HMA using artificial neural network. *Constr. Build. Mater.* **2018**, *170*, 527–536. [CrossRef]
34. Yusoff, N.I.M.; Breem, A.A.S.; Alattug, H.N.M.; Hamim, A.; Ahmad, J. The effects of moisture susceptibility and ageing conditions on nano-silica/polymer-modified asphalt mixtures. *Constr. Build. Mater.* **2014**, *72*, 139–147. [CrossRef]
35. Bala, N.; Napiah, M.; Kamaruddin, I. Effect of nanosilica particles on polypropylene polymer modified asphalt mixture performance. *Case stud. Constr. Mater.* **2018**, *8*, 447–454. [CrossRef]
36. Saltan, M.; Terzi, S.; Karahancer, S. Examination of hot mix asphalt and binder performance modified with nano silica. *Constr. Build. Mater.* **2017**, *156*, 976–984. [CrossRef]
37. Rezaei, S.; Ziari, H.; Nowbakht, S. Low temperature functional analysis of bitumen modified with composite of nano-SiO$_2$ and styrene butadiene styrene polymer. *Pet. Sci. Technol.* **2016**, *34*, 415–421. [CrossRef]
38. Adamu, M.; Mohammed, B.S.; Shafiq, N.; Liew, M.S. Durability performance of high volume fly ash roller compacted concrete pavement containing crumb rubber and nano silica. *Int. J. Pavement Eng.* **2018**, 1–8. [CrossRef]
39. Shafabakhsh, G.H.; Ani, O.J. Experimental investigation of effect of Nano TiO$_2$/SiO$_2$ modified bitumen on the rutting and fatigue performance of asphalt mixtures containing steel slag aggregates. *Constr. Build. Mater.* **2015**, *98*, 692–702. [CrossRef]
40. Leiva-Villacorta, F.; Vargas-Nordcbeck, A. Optimum content of nano-silica to ensure proper performance of an asphalt binder. *Road Mater. Pavement Des.* **2019**, *20*, 414–425. [CrossRef]
41. US Research Nanomaterials, Inc. Available online: https://www.us-nano.com/nanopowders?gclid=Cj0KCQjwg73kBRDVARIsAF-kEH-Yc3ah0ZJcGG8m4TlhRid-oo4L8wvoTMp_58Vv6s-9X_Tm530zdEQaAsZYEALw_wcB (accessed on 19 March 2019).
42. European Commission. Integrated pollution prevention and control (IPPC). Reference document on best available techniques for the manufacture of large volume inorganic chemicals—solids and others industry. 2007. Available online: http://eippcb.jrc.ec.europa.eu/reference/BREF/lvic-s_bref_0907.pdf (accessed on 12 December 2018).
43. Roes, A.L.; Tabak, L.B.; Shen, L.; Nieuwlaar, E.; Patel, M. Influence of using nanoobjects as filler on functionality-based energy use of nanocomposites. *J. Nanopart. Res.* **2010**, *12*, 2011–2028. [CrossRef]
44. Bare, J. TRACI 2.0: the tool for the reduction and assessment of chemical and other environmental impacts 2.0. *Clean Technol. Environ. Policy* **2011**, *13*, 687–696. [CrossRef]
45. Ryberg, M.; Vieira, M.D.; Zgola, M.; Bare, J.; Rosenbaum, R. Updated US and Canadian normalization factors for TRACI 2.1. *Clean Technol. Environ. Policy* **2014**, *16*, 329–339. [CrossRef]
46. Lippiatt, B.C. *BEES 2.0 Building for Environmental and Economic Sustainability: Technical Manual and User Guide*; National Institute of Standards and Technology: Gaithersburg, MD, USA, 2000.
47. Gladen, B.C.; Rogan, W.J. On graphing rate ratios. *Am. J. Epidemiol.* **1983**, *118*, 905–908. [CrossRef] [PubMed]

© 2019 by the authors. Licensee MDPI, Basel, Switzerland. This article is an open access article distributed under the terms and conditions of the Creative Commons Attribution (CC BY) license (http://creativecommons.org/licenses/by/4.0/).

MDPI
St. Alban-Anlage 66
4052 Basel
Switzerland
Tel. +41 61 683 77 34
Fax +41 61 302 89 18
www.mdpi.com

Applied Sciences Editorial Office
E-mail: applsci@mdpi.com
www.mdpi.com/journal/applsci